BLAST IN RAILROAD CUT AT ROSEVILLE MOUNTAIN, SUSSEX COUNTY, N. J.

Explosives In Metal Mining
And
Quarry Operations

By

Charles E. Munroe

And

Clarence Hall

ISBN 1-929148-29-1

First Printing 1915
Wexford College Press 2003

The Bureau of Mines, in carrying out one of the provisions of its organic act—to disseminate information concerning investigations made—prints a limited free edition of each of its publications.

When this edition is exhausted copies may be obtained at cost price only through the Superintendent of Documents, Government Printing Office, Washington, D. C., who is the authorized agent of the Federal Government for the sale of all publications.

The Superintendent of Documents *is not an official of the Bureau of Mines.* His is an entirely separate office, and he should be addressed:

SUPERINTENDENT OF DOCUMENTS,
Government Printing Office,
Washington, D. C.

The general law under which publications are distributed prohibits the giving of more than one copy of a publication to one person. Additional copies must be purchased from the Superintendent of Documents. The price of this bulletin is 25 cents.

First edition. January, 1915.

CONTENTS.

	Page.
Introduction	9
Combustion and explosion	10
Description of combustion	10
Description of explosion	10
Dust and gas explosions	10
Oxygen of air necessary to ordinary combustion	10
Oxidizing agents	11
Nitric acid supplies oxygen	11
Nitrocellulose	11
Nitroglycerin	12
Nitrosubstitution compounds	12
Reason for class names	12
Other explosive nitrogen compounds	13
Explosion by ignition	13
Explosion by detonation	13
Brisance	14
Efficiency of an explosive	14
Pressure of explosion exerted in all directions	14
Confining explosives	15
Danger from gaseous products of explosion	15
Example of poisoning in open quarry by gases from powder	16
Blasting and mine explosives	17
Low explosives and high explosives	17
Explosive compounds and explosive mixtures	17
Variety of explosives	17
Selection of an explosive for a particular project	18
Desirable characteristics of different explosives	18
Explosives used in quarrying	18
Black blasting powder	18
Volume and pressure of gases produced by explosives	19
Importance of time element	20
Cooling effect of drill holes	20
Rate of burning of blasting powder	20
Importance of stemming	20
Chambering drill holes	20
Granulated nitroglycerin powders	21
"Straight" nitroglycerin dynamites	21
Low-freezing dynamites	22
Ammonia dynamites	23
Gelatin dynamites	23
Gelatin dynamites used in wet blasting	24
Explosives used in tunneling and in mining work	24
Products of combustion	24
Poisonous gases from low-freezing dynamites	25

Page

Blasting and mine explosives—Continued.

Gas poisoning in tunnel _____ 25
Formula for a mine explosive _____ 26
Making a well-balanced formula _____ 23
Approved formula for gelatin dynamite _____ 27
Tests of approved gelatin dynamite _____ 27
Poisonous gases from gelatin dynamite improperly used _____ 28

Fuse, detonators, and electric detonators _____ 29

Methods of causing explosives to explode _____ 29
Fuse _____ 29
Advantages of Bickford's burning fuse _____ 29
Kinds of burning fuse on market _____ 30
Mode of action of burning fuse _____ 30
Standard rate of burning desirable _____ 30
Fuse should be tested for rate of burning _____ 31
Requirements burning fuse should satisfy _____ 31
Transportation and storage of burning fuse _____ 32
Use of burning fuse with detonators _____ 32
Classes of burning fuse _____ 32
Marking different fuse _____ 33
Care to be used in cutting fuse _____ 33
Detonating fuse _____ 34
Cordeau detonant _____ 34
Boosters _____ 35
Cordeau Bickford _____ 35
Detonators _____ 35
Advantages of using detonators in firing high explosives _____ 36
Grades of detonators _____ 36
Detonators weaker than No. 6 should not be used _____ 36
Method of attaching burning fuse to detonators _____ 37
Electric detonators _____ 37
Grades of electric detonators _____ 38
Firing electric detonators _____ 38
Clip for detonators _____ 39
Delay-action electric detonators _____ 39
Danger in use of detonators _____ 39
Storage and transporting of detonators _____ 40
Advantage of confining mercury fulminate _____ 40
Nail test for detonators _____ 40
Details of specific nail tests _____ 41
Failures from use of low-grade detonators _____ 42
Aged and insensitive explosives rendered efficient by use of high-
    grade detonators _____ 42
Tests with detonators distributed in charge _____ 42
Comments on the tests _____ 44
Simultaneous firing of several electric detonators in a charge _____ 44
Two kinds of explosives in the same drill hole _____ 44
Danger from use of combination charges in coal mines _____ 45
No advantage from combination charges in coal mines _____ 45
Ballistic-pendulum tests _____ 45
Results of tests _____ 46
No advantage from extra detonator in dynamite _____ 46
Electric igniters _____ 46

Page.

Firing blasts by electricity_____ 47
  Danger of firing by flame_____ 47
  Connecting legs to leading wires and making splices_____ 47
  Connecting leading wires to firing machine_____ 47
  Connecting detonators in series_____ 48
  Detonators (blasting caps) distributed in charge_____ 48
  Connecting holes_____ 48
  Danger from a break or short-circuit_____ 49
  Sources of current_____ 49
  Dry cells_____ 49
  Small firing devices are portable_____ 49
  Testing strength of batteries_____ 50
  Use of electric light or power circuit for firing blasts_____ 50
  Electric firing machines should not be overloaded_____ 50
  Proper manipulation of firing machine_____ 50
  Firing from electric lighting or power circuit_____ 51
  Dynamo-electric machines_____ 51
  Rating of dynamo-electric machines_____ 52
  Magnetos_____ 52
  Leading wires for firing machines_____ 52
  Inspection and repair of leading wires after firing_____ 52
  Testing the firing line_____ 52
  Safety precautions to be taken_____ 53
The use of explosives in excavation work_____ 54
  Efficiency explosive must be used_____ 54
  Improper selection costly_____ 54
  Various explosives for use in excavation work_____ 54
  Ammonia dynamites_____ 55
  Driving railroad cuts_____ 55
  Details of a large blast at a railroad cut_____ 55
    Preliminary excavation into the rock mass_____ 56
    Method of placing charge for the blast_____ 57
    Quantity and cost of materials_____ 57
    Miscellaneous details_____ 58
    Analyses of gaseous products of combustion_____ 59
    Results of blast_____ 59
  Submarine blasting_____ 59
  Advantages of "straight" nitroglycerin dynamite_____ 60
  Blasting in cities_____ 60
    Blasting a building site in Pittsburgh_____ 61
Use of explosives in quarrying_____ 63
  Conditions affecting choice of explosives_____ 63
  Increased use of broken rock_____ 63
  Stripping_____ 63
  Bench method of quarrying_____ 63
  Face method_____ 64
  Quarrying in Utah_____ 64
  Large blast in quarry at Tenino, Wash_____ 65
  Details of two large-scale quarrying operations_____ 68
    Quarrying at the Porto Bello quarry_____ 68
    Quarrying at the Ancon quarry_____ 69
  Blasts at Tomkins Cove, N. Y., quarry_____ 70

Page.

Use of explosives in metal mining and tunneling_____ 72
    Features of metal mines_____ 72
    Shaft sinking_____ 72
    Danger in blasting in shafts_____ 72
    Use of delay-action detonators_____ 73
    Driving tunnels_____ 73
    Use of detonators for firing dependent shots_____ 74
    Objection to fused detonators for wet blasting_____ 74
    Sealing with tallow_____ 75
    Use of new crimper_____ 75
    Use of explosives in large projects_____ 75
    New Buffalo waterworks tunnel_____ 75
        Size and character_____ 75
        Method of driving_____ 76
        Explosives used_____ 76
        Use of detonators_____ 77
        Handling of explosives_____ 77
        Firing shots_____ 77
    Drilling and blasting methods on New York rapid-transit tunnel (sub-
        way) _____ 78
    Rondout pressure tunnel_____ 78
    Hunter Brook Tunnel_____ 79
    The Elizabeth Tunnel_____ 80
        Size and character of tunnel_____ 80
        Method of drilling_____ 81
        Data on tunnel driving_____ 81
    Laramie-Poudre Tunnel _____ 82
        Size and character of tunnel_____ 82
        Tests of efficiency of explosives_____ 83
        Practice in use of explosives_____ 84
        Use of fuse_____ 84
        Ventilation _____ 85
        Drilling record_____ 85
    Data on hard-rock tunnelling in American tunnels_____ 86
    Trans-Andine Summit Tunnel_____ 86
    Simplon Tunnel _____ 87
    Loetschberg Tunnel_____ 89
Magazines and thaw houses_____ 94
    Storage and handling of explosives_____ 94
    Placing a magazine_____ 94
    Protection of explosives in magazines_____ 95
    Receiving explosives_____ 95
    Opening packages_____ 95
    Repair and care of magazines_____ 95
    Bullet-proof magazines_____ 96
    Lightning conductors_____ 96
    Use of galvanized-iron roof covering_____ 96
    Protection of life and adjacent property_____ 98
    Selection of magazine site_____ 98
    Combined table of distances_____ 98
    Barricades _____ 99
    Construction of magazines_____ 99

Page.

Magazines and thaw houses—Continued.

Bureau of Mines magazine_____ 100

Bill of materials_____ 100

Thawing explosives _____ 101

Precautions in thawing_____ 101

Thawing small quantities of explosives_____ 102

Thawing by manure_____ 102

Thawing large quantities of explosives_____ 103

Placing of thaw houses or magazines in cold climates_____ 103

Transporting thawed explosives_____ 103

Manure thaw houses_____ 103

Construction of thaw houses_____ 104

Bureau of Mines thaw house_____ 104

Bill of materials_____ 108

Hot-water heat in thaw houses_____ 109

Electric heaters_____ 109

Features of electric heating_____ 109

Comparison of heating methods_____ 109

Effect of heat changes_____ 110

Desirable capacity of heaters_____ 110

Miscellaneous essentials of heaters_____ 110

Permissible explosives_____ 112

Permissible explosives for coal mines_____ 112

Proposed requirements for metal-mine explosives_____ 112

Requirements for Bureau of Mines tests of metal-mine explosives_____ 114

Safe shipment and storage of explosives, by B. W. Dunn_____ 116

Responsibility of manufacturers and common carriers to public_____ 116

Federal law and Interstate Commerce Commission regulations_____ 116

Explosives in baggage or household goods_____ 116

Publications on mine accidents and tests of explosives_____ 118

ILLUSTRATIONS.

Page.

PLATE I. A, Explosion of high explosive in bomb filled with water; B, Screens for separating different-sized powder grains_____ 14

II. A, Wire screen and cartridges before firing; B, Wire screen after firing of cartridges_____ 14

III. A, Crarae quarry, Loch Fyne, Scotland; B, Bichel pressure gages and accessories_____ 16

IV. A, Black blasting powder, before sizing; B, Grains of black blasting powder _____ 20

V. A, Miner's squib; B, Miner's squib and fuse_____ 30

VI. A, X-ray print of defective fuse; B, Side spitting of a fuse____ 32

VII. A, Crimping detonator on fuse; B, Detonators and electric detonators _____ 38

VIII. A, Clip for detonator; B, Clip for electric detonator_____ 40

IX. A, New type of detonator box; B, Face of Ancon quarry_____ 40

Page.

PLATE X. *A*, Results of nail tests of P. T. S. S. electric detonators Nos. *3*,
4, 5, 6, 7, and 8; *B*, Results of nail tests of No. 6 electric
detonators _____   42

XI. *A*, Secondary battery, firing machine, and dry cell; *B*, Ballistic
pendulum_____   46

XII. *A*, Testing firing machines and batteries; *B*, Small electric lamp
used for testing batteries_____   50

XIII. Blast in railroad cut at Roseville Mountain, Sussex County,
N. J_____ Frontispiece

XIV. *A*, Ordinary crimper; *B*, New crimper_____   74

XV. *A*, Bureau of Mines cement-mortar magazine; *B*, Bureau of
Mines cement-mortar thaw house_____   96

FIGURE 1. Electric detonator, showing its component parts_____   37

2. Nail in position for test of electric detonator_____   41

3. Method of tunneling under large mass of rock before blasting_   56

4. Method of placing charge in tunnel for blasting large mass of
rock _____   57

5. Position of black-powder charges and of timber stemming in a
rock quarry at Tenino, Wash_____   66

6. Diagram of wiring at Tenino, Wash_____   67

7. Arrangement of drill holes in face and heading, new Buffalo
waterworks tunnel_____   76

8. Arrangement of drill holes in heading of Rondout Tunnel_____   79

9. Sections of Hunter Brook Tunnel, showing method of driving__   80

10. Arrangement of drill holes in Elizabeth Tunnel_____   81

11. Arrangement of drill holes in Laramie-Poudre Tunnel_____   83

12. Sections of Simplon Tunnel_____   _____   87

13. Sections of Loetschberg Tunnel_____   89

14. Section showing rocks penetrated by the Loetschberg Tunnel__   90

15. Plan and sections of Bureau of Mines cement-mortar magazine_   97

16. Roof plan and sections of Bureau of Mines cement-mortar
thaw house_____   106

17. Elevation and details of Bureau of Mines cement-mortar thaw
house_____   107

A PRIMER ON EXPLOSIVES FOR METAL MINERS AND QUARRYMEN.

By Charles E. Munroe and Clarence Hall.

INTRODUCTION.

In accidents resulting from the use of explosives in metal mines and quarries in the United States more than 130 men were killed and 250 seriously injured during the calendar year 1913. Moreover, an unknown number of miners suffered from the effects of breathing the harmful fumes and gases given off by the burning or the incomplete explosion of some explosive. Consequently, the Federal Bureau of Mines, which is endeavoring to increase safety in mines and to abolish conditions that tend to impair the health of miners, is studying the kinds of explosives used in mining and the conditions under which these explosives can be used with least danger to the miner.

Information about the different explosives used in coal mining and the precautions to be taken in using those explosives has been printed in Bulletin 17 and other bureau reports, the names of which may be found at the end of this bulletin. The bureau has not been able to give much attention to explosives used in metal mines and quarries, but has published some information about these explosives in Bulletin 48, "The Selection of Explosives Used in Engineering and Mining Operations"; Technical Paper 17, "The Effect of Stemming on the Efficiency of Explosives"; and Miners' Circular 19, "The Prevention of Accidents from Explosives in Metal Mining."

This bulletin aims to give the metal miner and the quarryman information similar to that given the coal miner in Bulletin 17. Inflammable gas or dust is seldom, if ever, found in quarries or metal mines, and the danger from using explosives there is less than in coal mines; but, as the figures show, the number of men killed and injured yearly in accidents caused by explosives proves the need of both miners and mine officials striving to see that none but proper explosives are used and that these are used properly.

9

COMBUSTION AND EXPLOSION.

DESCRIPTION OF COMBUSTION.

The action of explosives and the reason why they develop heat and power can be most easily understood by considering what takes place during ordinary burning or combustion. It is well known that when combustible substances—whether solids, such as wood, coal, charcoal, brimstone, or phosphorus; or liquids, such as petroleum, alcohol, or turpentine; or gases, such as coal gas or natural gas—burn with air, heat and light are emitted. After this burning or combustion, though some solid substances leave a varying amount of ash as a residue, the combustible part of the substances disappears from view by combining with the oxygen of the atmosphere to form gases. These gases at the instant of their formation are hot, and consequently are greatly expanded.

DESCRIPTION OF EXPLOSION.

Everybody who has kindled a fire knows that when a combustible solid is finely divided, as when wood is cut into shavings, it takes fire more readily. If a combustible substance be divided still finer—as when wood, charcoal, coal, or brimstone is crushed to powder—and then mixed with air and the mixture ignited, the rate of combustion is much more rapid and may become so great that an explosion follows.

DUST AND GAS EXPLOSIONS.

In fact, many terrible explosions have been caused by the ignition of combustible dusts suspended in air, not only in coal mines but in bitumen mines, sawmills, flour mills, starch mills, and elsewhere. Likewise, the vapors of combustible liquids—such as benzine, gasoline, or turpentine—when mixed with the air and ignited, have given rise to explosions, some of which have been almost appalling in magnitude. Combustible gases are still more likely to cause explosions, because gases form explosive mixtures with the air more readily than do powdered solids or sprayed or vaporous liquids. Many accidental explosions of such mixtures of combustible gases and air have been recorded.

OXYGEN OF AIR NECESSARY TO ORDINARY COMBUSTION.

In all these forms of combustion the combustibles burn with the oxygen in the air, and as a result of this burning the combustible elements unite with oxygen to form compounds, such as carbon monoxide (CO), carbon dioxide (CO_2), water vapor (H_2O), sulphur dioxide (SO_2), and still others, depending on the nature of the combustible element present.

OXIDIZING AGENTS.

. Centuries ago it was found that if such a combustible substance as charcoal, instead of being mixed with the air and ignited, was mixed with saltpeter or potassium nitrate (KNO_3), which was found as a coating on the soil in India, the mixture on ignition would burn out of contact with the air, the oxygen necessary for combustion being supplied by the saltpeter or nitrate. It was soon found that such a mixture was ignited with difficulty, and some ingenious person conceived the idea of adding to the mixture a little brimstone, which is easily ignited, so that the combustion of the brimstone might set fire to the mixture of saltpeter and charcoal. The sulphur, saltpeter, and charcoal burned rapidly, even when out of contact with air, and thus was gunpowder first made.

With the progress of the science of chemistry it has come to be known that there are many other solid substances that, like saltpeter, contain oxygen in a combined form and will, when in contact with a combustible substance, give up their oxygen on ignition, and thus enable the combustible substance to burn out of contact with the air. One of these substances is Chile saltpeter, or sodium nitrate ($NaNO_3$), which is found in large quantities in the desert of Tarapaca in Chile, and is to-day widely used in making explosives; others are potassium chlorate ($KClO_3$) and potassium perchlorate ($KClO_4$). Still other chlorates and perchlorates have been used in the compounding of explosives.

NITRIC ACID SUPPLIES OXYGEN.

Moreover, there are liquids in which oxygen is combined in an available form. Among such liquids, notably, is nitric acid (HNO_3), through the use of which it is possible to put combined oxygen in an available form more intimately in contact with combustible elements than can be done by mixing a pulverized solid oxidizing substance, such as saltpeter, with a pulverized solid combustible, such as charcoal.

NITROCELLULOSE.

It has been found, for instance, that if well-dried and well-purified cotton be dipped in nitric acid, or in a mixture of nitric acid and sulphuric acid (the latter being used to promote the reaction), what are known as nitryl groups—combinations of oxygen and nitrogen from the nitric acid, which have the chemical formula NO_2—can be substituted for one or more of the atoms of hydrogen in each molecule of the cellulose of the cotton. Investigation has further shown that when the nitryl groups are substituted for the hydrogen in the cellulose of cotton, the product is explosive. This explosive product is known as nitrocellulose, or, in some of its varieties, as guncotton, and is much used as a high explosive.

NITROGLYCERIN.

If, instead of cotton, the smooth, sweet, colorless liquid known as glycerin be used, a reaction takes place between it and nitric acid similar to that already described as taking place between cellulose and nitric acid. The atoms of hydrogen in the glycerin are replaced by nitryl groups, and the powerful explosive known as nitroglycerin is produced. Glycerin and cellulose are known to the chemists as alcohols. A study of the reactions indicated above showed that such bodies as starch, erythrol, mannitol, and numerous others contain alcoholic groups that give rise to similar explosive compounds, and these compounds are now being used in the manufacture of commercial explosives. All of these compounds contain nitryl groups, and so the arrangement of the atoms in their molecules resembles that of the potassium and sodium nitrates mentioned above.

NITROSUBSTITUTION COMPOUNDS.

In addition to the many alcohols, the chemist has also discovered the existence of a vast series of hydrocarbons, the first member of which, benzene (C_6H_6), obtained from coal tar, is the best known. When treated with nitric acid, each of these hydrocarbons, in common with benzene, forms explosive compounds in which one or more atoms of hydrogen in the hydrocarbon molecule are replaced by nitryl groups. Thus there are formed such substances as the nitrobenzenes, the nitrotoluenes, picric acid, and others that are known under the general name of nitrosubstitution compounds.

REASON FOR CLASS NAMES.

This name is given to distinguish these bodies from those derived from alcohols, as described above, which are known as nitric esters, and the distinction is made because in the nitrosubstitution compounds each nitryl group is attached directly to a carbon atom, whereas, in the nitric esters, such as nitrocellulose, nitroglycerin, and others, each nitryl or NO_2 group is attached indirectly, through the agency of an oxygen atom, to a carbon atom. This difference in naming is important not only because it indicates that the arrangement of the atoms in these different molecules is different, but also because it indicates that this difference in the arrangement of the atoms marks a difference in their behavior as explosives.

OTHER EXPLOSIVE NITROGEN COMPOUNDS.

Investigation has shown that there are other arrangements of atoms in molecules of nitrogen-containing substances. Such arrangements are found in the molecules of mercuric fulminate and other fulminates, of diazobenzene nitrate and other diazo compounds, of lead hydronitride and other salts of hydronitric or triazoic acid, and of other compounds. All of these compounds decompose readily by slight increase of temperature, by percussion, or by concussion, and when they are decomposed, by these or other means, they produce violent explosions; that is, the oxygen atoms unite with the carbon and hydrogen atoms of the molecules so swiftly that the energy stored in the molecules is set free instantly in so concentrated a condition that it can set off explosives that are not readily exploded by an ordinary shock or by a decided increase of temperature.

EXPLOSION BY IGNITION.

It has been pointed out that by introducing the nitryl, or nitro, group into certain kinds of molecules to form the nitric esters and nitrosubstitution compounds there are formed explosive compounds that, like guncotton, can be exploded more or less readily by ignition, because the oxygen atoms are intimately associated with the hydrogen and carbon atoms in each molecule. These compounds, whether solid or liquid, on explosion form gaseous products that have a volume much greater than that of the solid or liquid, and as these gaseous products are hot, and consequently tend to expand with force, they are, if confined, capable of doing such work as bursting masses of rock or driving projectiles from cannon.

EXPLOSION BY DETONATION.

It has been found that these substances, particularly the nitric esters in which the molecules can be shaken apart, can be exploded in other and much more effective ways than by simple ignition. One of these ways is by the use of an exploding device, such as a blasting cap, or detonator, containing mercury fulminate or the hydronitride of a metal; another is by the explosion of a contiguous mass of the same ester. This method of enabling the oxygen atoms to react with other atoms in the molecules of the explosive is styled detonation, and when the nitric esters or the nitrosubstitution compounds, or mixtures thereof with other substances, are exploded by detonation the explosive change is much more rapid than the change produced by ignition, so that the solid or liquid explosives are converted into highly heated and greatly expanded gases much more quickly. Therefore when exploded by detonation the effect of the explosive, especially in breaking surrounding material, is greatly increased.

BRISANCE.

This quick or shattering effect caused by the high velocity of the explosive reaction is termed "brisance," and is clearly much different from the effect produced by the explosion of slow-burning gunpowder or black blasting powder. Hence explosives have come to be classified as low explosives and high explosives. Gunpowder and black blasting powder are typical low explosives, and nitroglycerin and the dynamites made from it are typical high explosives.

EFFICIENCY OF AN EXPLOSIVE.

The efficiency of an explosive may be judged by various standards, but the user of an explosive is most concerned with the ability of the explosive to do the work desired. This ability is measured by both the quality and the quantity of the products that it yields. The effects vary not only with the kind of explosive, the manner in which it is used, and the method by which it is exploded, but also with the character of the rock or other material in which it is used. All other characteristics of certain explosives under consideration being the same, the rate of detonation becomes the governing factor in determining the most efficient, and affords the best means for selecting explosives suitable to meet the varying conditions in mining, quarrying, and blasting operations.

PRESSURE OF EXPLOSION EXERTED IN ALL DIRECTIONS.

Every explosive when exploded exerts pressure in every direction. The fact that an explosive does not exert pressure downward rather than upward is shown by the results of two experiments. In one a naval detonator, filled with mercuric fulminate, was exploded in a closed iron cylinder filled with water. The cylinder was blown into the form of a sphere and ruptured. (See Pl. I, *A*.) In the other a cartridge of dynamite was placed on one side of a wire screen suspended vertically in the air and a similar cartridge of dynamite was placed on the other side of the screen at a distance from the first cartridge. (Pl. II, *A*.) Both were exploded simultaneously, with the result, as shown in Plate II, *B*, that holes were cut through the screen in opposite directions. But, when laid on top of a rock and exploded, gunpowder and other low explosives do not materially

A. EXPLOSION OF HIGH EXPLOSIVE IN BOMB FILLED WITH WATER.

B. SCREENS FOR SEPARATING DIFFERENT-SIZED POWDER GRAINS.

A. WIRE SCREEN AND CARTRIDGES BEFORE FIRING.

Wire screen of No. 10 wire is suspended vertically in the air with two ½-pound cartridges of 40 per cent "straight" dynamite wired against it.

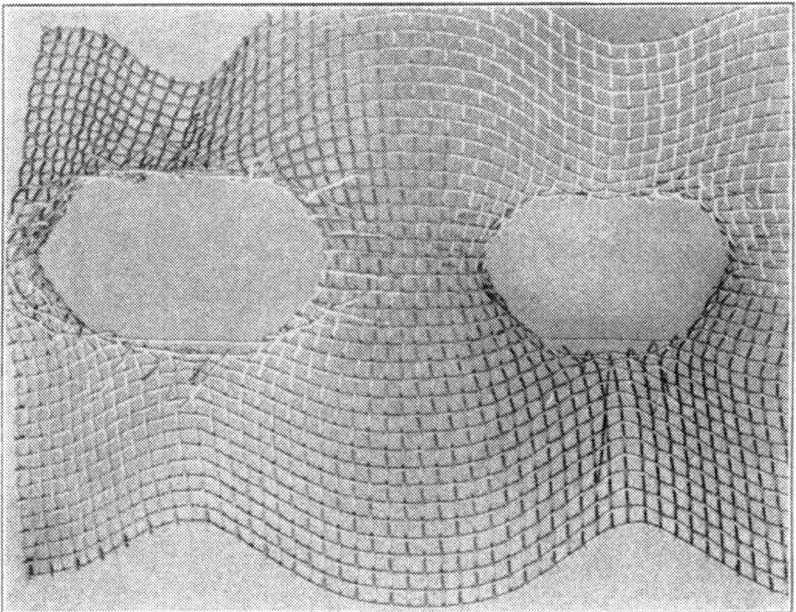

B. WIRE SCREEN AFTER FIRING OF CARTRIDGES.

affect the rock, because they explode so slowly that the gases formed can lift the air above them and escape; yet dynamite, fulminate of mercury, and other high explosives, if laid on brittle or soft rock and detonated, may shatter it, because they explode so quickly that the gases formed from them can not lift the large volume of air confining them without pressing back forcibly against the rock.

CONFINING EXPLOSIVES.

This confinement by air is not, however, close enough to give the best result with any explosive. By boring a hole in rock and confining the explosive firmly in it by means of well-tamped stemming, gunpowder and other low explosives may be made to break the rock, or a much less quantity of high explosive may be used than is required to break it when the explosive is merely laid on it. Confining an explosive is the cheapest and best way of using it so as to obtain the most service from it.

DANGER FROM GASEOUS PRODUCTS OF EXPLOSION.

It should always be remembered that the gases produced by explosives in exploding have other properties than that of yielding larger volumes than the solids from which they are produced or of expanding when heated, and thus being capable of doing work, for some of these gases are combustible and may form explosive mixtures with the air and others are poisonous. The kind of gases produced differs with the kind of explosives used and the way in which it is used. Nitric esters and nitrosubstitution compounds and the mixtures, such as dynamite, made from them may yield poisonous nitrogen oxides and cyanogen, and poisonous and inflammable hydrogen sulphide and carbon monoxide. But if the explosive be properly compounded, well confined, and fully detonated, these harmful gases will not be produced, for the gaseous products will then be largely composed of water and carbon dioxide; and though carbon dioxide may cause unconsciousness and even death, it does so only when it forms a large proportion of the atmosphere. Blasting powder and gunpowder-like mixtures give off poisonous and inflammable hydrogen sulphide and carbon monoxide under all conditions of explosion.

The production of inflammable gaseous products underground, especially in coal mines and bitumen mines, is most hazardous. The production of poisonous gases, either below ground or on the surface, is a source of danger, and if such gases are formed they should be removed by ventilation before anyone is allowed to approach the working place.

EXAMPLE OF POISONING IN OPEN QUARRY BY GASES FROM POWDER.

Neglect of this precaution led to the death of 7 persons and the rendering of 40 others unconscious from gas poisoning following the firing of 21,000 pounds of blasting gunpowder at the Crarae quarry, Loch Fyne, Scotland, on September 25, 1886. The quarry was situated in a basin in a hill with sides rising 25 to 250 feet, and was approached by a narrow gorge. (See Pl. III, A.) The blast was fired in the presence of an audience of over 1,000 persons stationed on a steamer about a mile distant. At least half an hour elapsed after the blast had been fired before passengers from the steamer, to the approximate number of 120, got into the quarry to observe the effect, and within five to six minutes after entrance they began to fall, overcome by the poisonous powder gases entangled in the crevices of the rock that had been thrown down.

A. CRARAE QUARRY, LOCH FYNE, SCOTLAND.

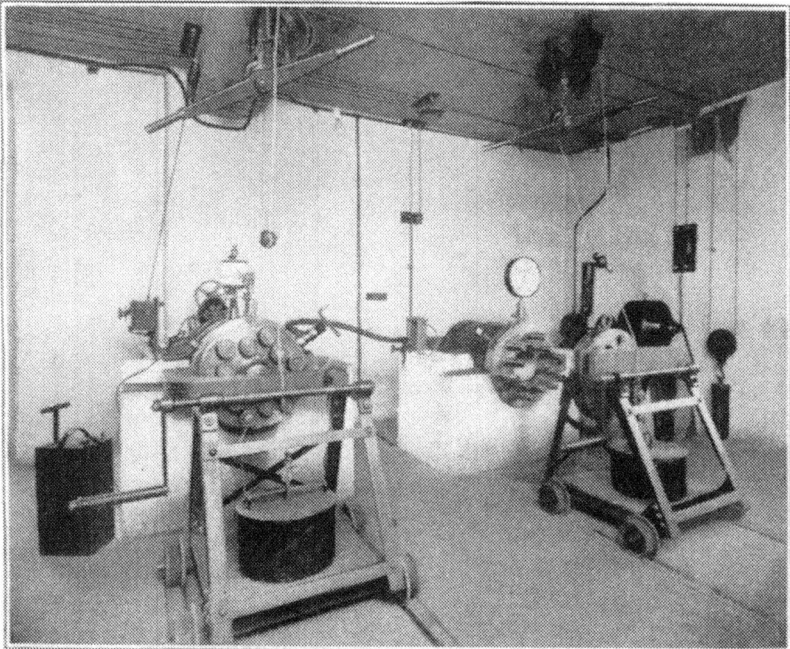

B. BICHEL PRESSURE GAGES AND ACCESSORIES.

BLASTING AND MINE EXPLOSIVES.

LOW EXPLOSIVES AND HIGH EXPLOSIVES.

Explosives have already been classified, according to their action, as low explosives and high explosives. The chief characteristic of low explosives is that they produce on explosion a propulsive or pushing effect, and of high explosives that on explosion they produce mainly a disruptive or shattering effect.

EXPLOSIVE COMPOUNDS AND EXPLOSIVE MIXTURES.

Other schemes of classification of explosives are used, a notable one being based on their composition. Under this classification substances like nitroglycerin and guncotton are styled nitric esters, and substances like nitrotoluene and nitronaphthalene are styled nitro-substitution products. All of these, with the fulminates and hydro-nitrides, are definite chemical compounds, whereas gunpowder, black blasting powder, dynamite, and numerous other explosive compositions are mixtures and may contain both chemical compounds and elementary substances. Therefore explosives may also be classified as explosive compounds and explosive mixtures. Even the explosive compounds, such as nitroglycerin and the nitrosubstitution products, are not commonly used by themselves, but as components of a large number of explosive mixtures, such as the various dynamites.

VARIETY OF EXPLOSIVES.

It is clear from what has been said that a large number of different explosive compositions may be made, and a large number are made and are placed on the market for sale and use. As each different explosive compound has different characteristics, it is obvious that each different explosive mixture made with the different compounds has different characteristics, according to the proportion of a given explosive compound and of the components with which it is mixed, and that some of these explosive mixtures are better adapted for use under a particular set of conditions than are others.

SELECTION OF AN EXPLOSIVE FOR A PARTICULAR PROJECT.

In large engineering projects and in mining and quarrying operations in which the work is directly dependent on the use of explosives, the selection of a suitable explosive from the many varieties offered for sale is of fundamental importance. Many considerations are involved in the selection of the proper class of explosives for use in such operations, because most explosives suitable for use in such work as quarrying are unsuited for use in deep mines or in close workings.

DESIRABLE CHARACTERISTICS OF DIFFERENT EXPLOSIVES.

In wet workings or in submarine blasting, the cartridges of explosives must not be affected by moisture. In extremely cold climates explosives that do not require thawing are desirable. In metal mining and in driving tunnels the character of the gases evolved by the explosive on detonation is one of the important considerations in determining the selection. Explosives for use in gaseous or dusty coal mines must be so compounded that their flame temperatures and the length and duration of their flames are so moderate that they can be used in such mines with comparative safety. An important requirement of all explosives, especially those for use in tropical countries, is that they shall remain stable and shall not be affected by storage for a reasonable length of time.

EXPLOSIVES USED IN QUARRYING.

In order to meet the varying conditions in open work, such as quarrying, practically every class and grade of commercial explosive is used. In breaking down rock in quarries the explosives in general use are: Black blasting powder; granulated powder, containing 5 to 15 per cent of nitroglycerin; "straight" nitroglycerin dynamites, varying in grade from 15 to 60 per cent of nitroglycerin; low-freezing dynamites, rated, according to their equivalent percentages of strength, with the "straight" nitroglycerin dynamites; and ammonia dynamites, rated in a similar way. In recent years blasting gelatin, gelatin dynamites, ammonium nitrate powders containing nitro-substitution compounds, chlorate powders, and the so-called nitrostarch powders have been introduced in a small way in quarry operations.

BLACK BLASTING POWDER.

Black blasting powder is a mechanical mixture; it consists of finely pulverized sodium nitrate intimately associated with combustible materials. This intimate association of the materials is produced by thoroughly mixing them in what are known as wheel mills. The product is pressed into hard cakes, which are subsequently cracked into small grains. The edges of the grains are rounded by causing them to be rubbed together in a revolving barrel, and at the same time the surfaces are coated or glazed, as it is termed, with graphite, often called black lead. The object of rounding and glazing the grains is to render them free running; the glazing also serves to hinder their taking up moisture from the air, because of their containing the deliquescent or moisture-absorbing substance, sodium nitrate. Black blasting powder, such as is generally offered for sale in this country, has the following composition:

Composition of black blasting powder.

	Per cent.
Charcoal	16
Sulphur	11
Sodium nitrate	73
	100

The powder (Pl. IV, *A*) is divided into grades according to size of grains, as separated and collected by sieves of different sizes of openings. (Pl. I, *B*.) The sizes most usually offered for sale are called CC, C, F, FF, FFF, and FFFF. (Pl. IV, *B*.) Of these, CC represents the largest grains, which are half an inch in diameter, and FFFF represents the smallest grains, which are a sixteenth of an inch in diameter.

VOLUME AND PRESSURE OF GASES PRODUCED BY EXPLOSIVES.

When blasting powder is ignited in a drill hole a rapid combustion takes place and the pressure by the highly heated gases forces apart or blasts the rock or other material. It has been found that 150 liters (5.30 cubic feet) of gas measured at normal temperature (0° C. or 32° F.) and pressure (760 mm., or 30 inches, of mercury) is produced from 454 grams (1 pound) of black blasting powder. According to Berthelot, the famous French chemist, nitroglycerin yields three and one-half times as great a volume of permanent gases as does the same weight of nitrate powder. When these volumes are compared with the relative effectiveness of the two explosives it is evident that other factors quite as important as the volume of permanent gases evolved must be taken into account in comparing the energy developed by different explosives. Among these factors [a] is the heat set free in the explosion. The theoretical temperature produced at the instant of explosion by dynamite is much higher than that produced by the explosion of black blasting powder, and consequently the gases evolved on the detonation of dynamite are more highly heated and expanded than those resulting from the explosion of black blasting powder.

---

[a] See Hall, Clarence, Snelling, W. O., and Howell, S. P., Investigation of explosives used in coal mines: Bull. 15, Bureau of Mines, 1912, pp. 16–62.

IMPORTANCE OF TIME ELEMENT.

In determining the efficiency of commercial explosives the rate of detonation or burning is the governing factor and offers the best means for selecting explosives suitable to meet the varying conditions of mining work.

COOLING EFFECT OF DRILL HOLES.

During the conversion of an explosive into solid, liquid, and gaseous products the cooling effect of the walls of the drill hole or other heat-absorbing surroundings tends to reduce the temperature of the gases so that neither the theoretical temperature nor the maximum pressure is ever reached. The more nearly instantaneous the explosive reaction of any explosive, all other conditions being equal, the greater will be the volume of highly heated gases produced and the more violent the effect.

RATE OF BURNING OF BLASTING POWDER.

According to experiments made at the Pittsburgh experiment station of the Bureau of Mines the rate of burning of FF black blasting powder under conditions resembling those present when the powder is exploded in a drill hole is 469 meters (1,538 feet) per second. This rate is relatively slow as compared with the rate of detonation of "ordinary" dynamites, and hence the gases from the explosion of black blasting powder, being evolved much more slowly and having more time to lose heat and pressure by conduction and radiation, give pressures that are prolonged. Hence black blasting powder is best suited for work in which a gradual pushing or heaving effect is desired.

IMPORTANCE OF STEMMING.

In order to obtain the maximum efficiency with black blasting powder the charge must be well confined by suitable stemming.[a] The best results are obtained when the drill holes are tamped to the mouth.

CHAMBERING DRILL HOLES.

Black blasting powder is used in excavating earth cuts, in strippings, and in quarrying soft rock or stone, especially in quarries where dimension stone is sought, if an explosive must be used there.

---

[a] See Snelling, W. O., and Hall, Clarence, The effect of stemming on the efficiency of explosives: Technical Paper 17, Bureau of Mines, 1912, 20 pp.

A. BLACK BLASTING POWDER, BEFORE SIZING.

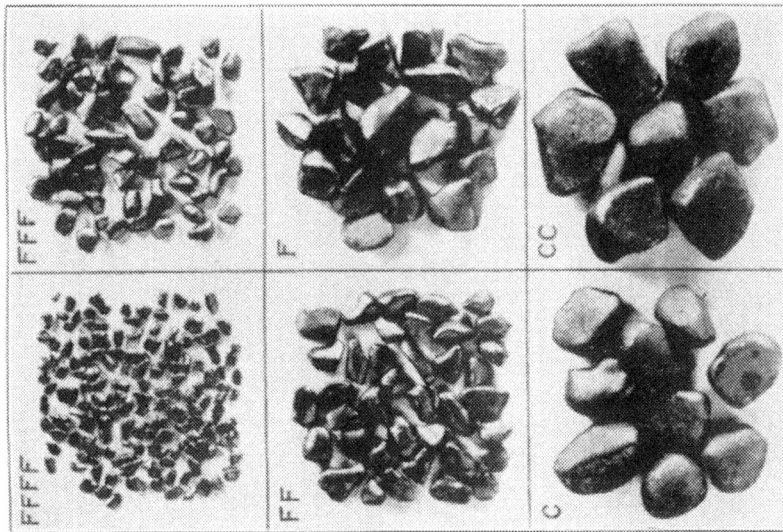

B. GRAINS OF BLACK BLASTING POWDER.

In some work, such as in winning broken stone for concrete or ballast and in open-cut operations, it is necessary to chamber the bottom of the drill hole with dynamite or other suitable high explosive before charging it with black blasting powder. By this procedure a large cavity is formed at that point where the powder can be placed to the best advantage. When drill holes are chambered they become hot and should not be reloaded immediately after a chambering shot has been fired, for the new charge may explode prematurely.

GRANULATED NITROGLYCERIN POWDERS.

Granulated nitroglycerin powder such as is generally offered for sale in this country has about the following composition:

Composition of granulated nitroglycerin powder.

	Per cent.
Nitroglycerin	5
Combustible material [a]	35
Sodium nitrate	60
	100

The composition of the grains of granulated nitroglycerin powders is somewhat similar to that of the grains of black blasting powder, but instead of the nitrate and combustible materials being thoroughly mixed in a wheel mill the grains of these powders are made by heating the ingredients of the mixture to a temperature at which the resin and sulphur melt, and the sticky mass thus formed is then granulated while cooling by being rubbed through a fine-mesh screen. The grains produced in this way are hard and porous, and when cool they are mixed with a small percentage of nitroglycerin, which partly coats their surfaces and partly clings in their pores.

To obtain the maximum efficiency from a charge of granulated powder it is necessary to detonate the charge with a priming cartridge of dynamite. The priming cartridge detonates the nitroglycerin held in the pores and on the outside of the grains, and the combustion of the powder grains is thus hastened. The rate of detonation of granulated powder is 1,018 meters (3,339 feet) per second, or more than twice that of the burning of black blasting powder. Thus granulated powders, when used in soft, seamy rock, are more effective and give better results than does black blasting powder.

"STRAIGHT" NITROGLYCERIN DYNAMITES.

Compositions.—The following compositions are examples of "straight" nitroglycerin dynamites such as are generally offered for sale in this country:

---

[a] Consists of sulphur, coal, and resin.

Compositions of typical " straight " nitroglycerin dynamites.

Ingredient.	15 per cent strength.	20 per cent strength.	25 per cent strength.	30 per cent strength.	35 per cent strength.	40 per cent strength.	45 per cent strength.	50 per cent strength.	55 per cent strength.	60 per cent strength.
Nitroglycerin	15	20	25	30	35	40	45	50	55	60
Combustible material [1]	20	19	18	17	16	15	14	14	15	16
Sodium nitrate	64	60	56	52	48	44	40	35	29	23
Calcium or magnesium carbonates	1	1	1	1	1	1	1	1	1	1
	100	100	100	100	100	100	100	100	100	100

[1] Consists of wood pulp, flour, and brimstone for grades below "40 per cent"; other grades, wood pulp only.

Rate of detonation.—The rates of detonation [a] of 30 per cent and 60 per cent "straight" nitroglycerin dynamites have been found at the Pittsburgh experiment station of the Bureau of Mines to be 4,548 meters (14,920 feet) and 6,246 meters (20,490 feet) per second.

Disruptive force.—The "straight" nitroglycerin dynamites, as a class, have also been found there to develop greater disruptive force than any of the other commercial types of explosives tested, and for this reason they should be used for producing shattering effects or in blasting very tough or hard materials, whenever the conditions permit.

LOW-FREEZING DYNAMITES.

If "straight" nitroglycerin dynamites prove to be too violent for certain classes of work, the low-freezing dynamites or the ammonia dynamites, which have lower rates of detonation and exert more nearly a simple propulsive or pushing action, may be used to advantage.

The following table shows typical compositions of low-freezing dynamites:

Compositions of typical low-freezing dynamites.

Ingredient.	30 per cent strength.	35 per cent strength.	40 per cent strength.	45 per cent strength.	50 per cent strength.	55 per cent strength.	60 per cent strength.
Nitroglycerin	23	26	30	34	38	41	45
Nitrosubstitution compounds	7	9	10	11	12	14	15
Combustible material [1]	17	16	15	14	14	15	16
Sodium nitrate	52	48	41	40	35	29	23
Calcium or magnesium carbonate	1	1	1	1	1	1	1
	100	100	100	100	100	100	100

[1] Composition similar to that of the combustible material in the "straight" nitroglycerin dynamites.

The low-freezing dynamites have the advantage of not freezing until a temperature of 35° F. or less is reached, but, like all nitroglycerin explosives, when they become frozen they must be thawed before use in order to insure their giving the most effective results.

[a] For method of measurement see Hall, Clarence, and others, Investigations of explosives used in coal mines: Bull. 15, Bureau of Mines, 1912, pp. 92–95.

AMMONIA DYNAMITES.

The following tables show compositions of ammonia dynamites:

Compositions of typical ammonia dynamites.

Ingredient.	30 per cent strength.	35 per cent strength.	40 per cent strength.	50 per cent strength.	60 per cent strength.
Nitroglycerin	15	20	22	27	35
Ammonium nitrate	15	15	20	25	30
Sodium nitrate	51	48	42	36	24
Combustible material a	18	16	15	11	10
Salcium carbonate or zinc oxide	1	1	1	1	1
	100	100	100	100	100

a Composition similar to that of the combustible material in the "straight" nitroglycerin dynamites of grades below 40 per cent strength.

Compositions of low-freezing ammonia dynamites of various strengths.

Ingredient.	30 per cent strength.	35 per cent strength.	40 per cent strength.	50 per cent strength.	60 per cent strength.
Nitroglycerin	13	17	17	21	27
Nitrosubstitution compounds	3	4	4	5	6
Ammonium nitrate	15	15	20	25	30
Sodium nitrate	53	49	45	36	27
Combustible material a	15	14	13	12	9
Calcium carbonate or zinc oxide	1	1	1	1	1
	100	100	100	100	100

a Composition similar to that of the combustible material in the "straight" nitroglycerin dynamites of grades below 40 per cent strength.

As the ammonia dynamites, when compared with the other dynamites, have the disadvantage of readily taking up moisture, because the ammonium nitrate in them absorbs moisture from the air, care should be taken not to store or use them in wet places.

GELATIN DYNAMITES.

The following compositions are examples of gelatin dynamites generally offered for sale in this country:

Compositions of typical gelatin dynamites.

Ingredient.	30 per cent strength.	35 per cent strength.	40 per cent strength.	50 per cent strength.	55 per cent strength.	60 per cent strength.	70 per cent strength.
Nitroglycerin	23.0	28.0	33.0	42.0	46.0	50.0	60.0
Nitrocellulose	0.7	0.9	1.0	1.5	1.7	1.9	2.4
Sodium nitrate	62.3	58.1	52.0	45.5	42.3	38.1	29.6
Combustible material a	13.0	12.0	13.0	10.0	9.0	9.0	7.0
Calcium carbonate	1.0	1.0	1.0	1.0	1.0	1.0	1.0
	100.0	100.0	100.0	100.0	100.0	100.0	100.0

a Wood pulp used in 60 and 70 per cent strengths. Flour, wood pulp, and, in some makes, resin and brimstone used in other grades.

Some manufacturers replace a small percentage of the nitro-glycerin in these grades with an equal proportion of ammonium nitrate, but the substitution offers little, if any, advantage other than to reduce the cost of manufacture.

GELATIN DYNAMITES USED IN WET BLASTING.

In wet blasting the gelatin dynamites have been much used for removing obstacles to navigation, and they have been used in deep workings, and, as a general rule, are best adapted for these purposes. In the process of manufacturing gelatin dynamites the nitro-glycerin is first gelatinized by the addition of a small percentage of nitrocellulose. In this way a jelly-like mass, known as explosive gelatin, is formed, which is impervious to water. By mixing this mass with different proportions of suitable materials the various grades of gelatin dynamites are made.

EXPLOSIVES USED IN TUNNELING AND IN MINING WORK.

It is generally recognized that in driving tunnels, in sinking shafts, and in metal mining, where hard rock must be blasted, explosives having a high disruptive or shattering force are preferable. In such work the cost of drilling holes is of great importance, and it has been found more economical to drill a few holes and load them with an explosive having a high disruptive force than to drill a large number of holes and to use a weaker and cheaper explosive. It is important that explosives used in this class of work should produce the minimum amount of poisonous gases.

From what has been said regarding the composition and the characteristics of the different classes of explosives, it is evident that black blasting powder, granulated nitroglycerin powder, low-freezing dynamites, and the ammonia dynamites would not be suitable or satisfactory for blasting hard rock, because they do not develop enough disruptive force; furthermore, these explosives, in common with the "straight" nitroglycerin dynamites, produce considerable quantities of poisonous gases, which vitiate the atmosphere of the working places.

PRODUCTS OF COMBUSTION.

At the Pittsburgh experiment station of the bureau Bichel pressure gages (Pl. III, *B*) are used to determine the maximum pressures developed by explosives, and the cylinders of the gages are also used for collecting the gaseous, liquid, and solid products of explosion. The gages comprise two stout steel cylinders that may be made air-tight, each having an insulated plug for providing a means of igniting the charge, an air pump and proper connections for exhausting the air in either cylinder, a valve by means of which this air pump can be isolated from the cylinder, and an indicator mechanism, the drum of which may be driven by a motor at a known speed. The cover of each cylinder is a heavy piece of steel held in place by 12 heavy stud bolts and an iron yoke. The movement of the piston of the indicator is resisted by one of several spiral springs of different strength. The particular one used is determined by the character of the explosive to be tested. The pressure developed within the cylinder is recorded by means of a stylus on a paper placed on the circumference of the indicator drum.

In seeking to collect the gaseous and other products of explosion the same method is used in preparing and firing the charge as is followed in determining the maximum pressures developed, except that the indicator mechanism used with the cylinders is replaced by a pressure gage. After the charge has been fired the gases are allowed to cool for 30 minutes, the pressure is recorded, and the volume of gases at 760 mm. (30 inches) and 0° C. (32° F.) is computed. A sample of the gases is then taken over mercury by allowing the gases to escape gradually from the gage. A charge of 200 grams of the explosive, in its original wrapper, is used in all tests, except with those of black blasting powder, which is fired in charges of 300 grams.

POISONOUS GASES FROM LOW-FREEZING DYNAMITES.

The low-freezing dynamites produced on detonation larger volumes of poisonous gases than any other explosive tested, one sample of the gaseous products of combustion of a low-freezing dynamite containing 47.47 per cent by volume of carbon monoxide. The "straight" nitroglycerin dynamites produced 26.9 to 34.6 per cent of carbon monoxide; black blasting powder produced 10.8 per cent carbon monoxide and 8.7 per cent hydrogen sulphide; granulated powder produced 2.7 per cent carbon monoxide and 15.7 per cent hydrogen sulphide; and ammonia dynamite produced 3.8 per cent monoxide and 5.4 per cent hydrogen sulphide. The gelatin dynamites on detonation gave the least amount of poisonous gases, namely, 3.0 per cent carbon monoxide and 4.1 per cent hydrogen sulphide. This amount is yet far from being satisfactory; nevertheless, the gelatin dynamites are much used in this country for blasting in metal mines and in tunnels.

GAS POISONING IN TUNNEL.

In many metal mines the ventilation is defective or insufficient and, when the work has required the use of large quantities of explosives, many accidents have been caused by the poisonous gases evolved. In a large engineering project in the West nine men lost their lives as a result of the poisonous gases produced on the detonation of 40 per cent strength gelatin dynamite in a long tunnel.

After igniting the blast the men retired about 500 feet to wait for the smoke to clear, and while they were waiting the smoke drifted slowly over them, and then, owing to some change in the air current, drifted slowly back again. The men felt the usual symptoms of carbon monoxide poisoning—slight choking, nausea, profuse perspiration, and headache—but they all revived upon reaching the open air about an hour and a half after the blast had been fired. Within a short time, however, the men began to cough and spit bloody mucus and show other symptoms of nitrogen peroxide poisoning. In less than three days 9 out of the 13 men who had been in the tunnel and exposed to the fumes had died; the other 4, as well as those who went in with the motor to bring the men out, were ill for days and even months after the catastrophe.

It was soon after the accident mentioned above that special studies of the noxious gases evolved on the detonation or combustion of different explosives were undertaken by the Bureau of Mines to determine whether improvements could be made in the composition of explosives with a view to increasing safety in mining.

FORMULA FOR A MINE EXPLOSIVE.

From what has been said it is evident that there are available only two classes of explosives that develop the disruptive force required for tunneling and mining in hard rock; namely, "straight" nitroglycerin dynamite and gelatin dynamite. Well-balanced formulas for the "straight" nitroglycerin dynamites can be made by reducing the proportion of combustible materials in the "dope," and thus eliminating the oxygen deficiency, by replacing the wood pulp in the higher grades with inactive material, like magnesium carbonate, or kieselguhr (infusorial earth), having a high absorbing capacity. However, as decreasing the proportion of active material decreases the efficiency of an explosive, it was thought advisable to conduct experiments with gelatin dynamites, which do not require combustible material for absorbing and retaining the nitroglycerin present in them.

Several samples of 40 per cent strength gelatin dynamite were procured from different manufacturers, and on detonation all samples produced poisonous gases, the percentage of carbon monoxide varying from 3 to 5.7 per cent.

MAKING A WELL-BALANCED FORMULA.

Inspection of the formulas for gelatin dynamite given on page 23 shows that there is seemingly oxygen enough in the sodium nitrate to oxidize completely the combustible materials, and for this reason all the formulas are termed well balanced. However, in the gage tests hydrogen sulphide was evolved and remained in the permanent gases resulting from detonation. It was concluded that the brimstone used in the gelatin dynamites was responsible for the poisonous gas, and should not be used as an ingredient. Further investigations showed that the paper wrappers used in connection with the explosive, which were heavily coated with paraffin and represented a comparatively large percentage by weight of the combustible material present, had not been given due consideration by the manufacturers of explosives. It was believed that by slightly increasing the percentage of sodium nitrate sufficient oxygen would be provided to completely oxidize the hydrogen and carbon contents of the paper wrapper.

APPROVED FORMULA FOR GELATIN DYNAMITE.

As a result of the investigations a quantity of 40 per cent strength gelatin dynamite was manufactured according to the following formula:

Formula for gelatin dynamite.

	Per cent.
Nitroglycerin	33
Nitrocellulose	1
Sodium nitrate	54
Combustible material [a]	11
Calcium carbonate	1
	100

TESTS OF APPROVED GELATIN DYNAMITE.

On detonation in the gages the 40 per cent strength gelatin dynamite made according to this modified formula did not evolve poisonous gases. The products of combustion were determined in the gage, 200 grams of the explosive in the original wrapper being used under the same conditions as in previous tests. However, as it was feared that poisonous gases might be evolved when large quantities were detonated under actual mining conditions, the gage tests were supplemented by tests in mines. The first of these experiments were made in a limestone mine at West Winfield, Pa. In this mine the limestone was mined by the room-and-pillar system, and the limestone was shot "off the solid," four or more shots being used, according to the width of the face to be blasted. Hence different amounts of explosives had to be used in the different working places. No. 6 electric detonators were used, and all the shots in a face were connected in series and fired at one time.

[a] Flour.

The second series of experiments was made in a zinc mine at Franklin Furnace, N. J. The headings in which the tests were made averaged about 5 feet wide and 7 feet high. In driving these headings a 21-hole cut was used. The first shot, which was fired in the center of the face, was of the type known as a "burning" shot. Such a shot is fired, not with the object of blasting out any material, but only to form a crack in the center of the face or to so weaken the rock about the blast hole that when the first round of shots is fired the material can easily be blasted out. The method of loading and firing the second, third, and fourth rounds of shots was similar to that generally followed in metal-mining work. The samples were taken by air displacement by means of a small bellows which was emptied and filled 50 times. In taking some of the mine-air samples it was deemed necessary to use a breathing apparatus.

Although limited in number and extent, the results of the field tests, with a few exceptions, confirmed the tests made in the pressure gage. The odor of hydrogen sulphide was noticeable immediately after some of the explosives containing brimstone as an ingredient had been fired, but the chemical analyses of the mine-air samples failed to disclose the presence of any appreciable amount of this gas. Several days intervened between the taking of the samples in the mine and the chemical examinations, and if there were minute quantities of hydrogen sulphide present in any of the samples the hydrogen sulphide may have been altered or otherwise changed by reason of the samples standing so long. It is worthy of note that in all tests made the explosives were completely detonated, and there was no formation of nitrogen oxides.

POISONOUS GASES FROM GELATIN DYNAMITE IMPROPERLY USED.

The results of all the experiments indicate that all gelatin dynamites should be made with a sufficient oxygen excess to completely oxidize all combustible materials present, including the paraffin-coated paper wrappers; furthermore, the tests show that when this type of explosive is properly made and completely detonated the amount of harmful gases evolved is reduced to a minimum. This freedom from poisonous gases will not exist if there has been any chemical or physical change in the explosive, or if it is fired under conditions that would cause an incomplete detonation or burning, for under such conditions a great quantity of poisonous gases will be evolved. These conditions will exist when the explosive has aged to such an extent as to materially decrease its sensitiveness, when weak detonators are used, or when the explosive is used in a frozen or partly frozen condition.

FUSE, DETONATORS, AND ELECTRIC DETONATORS.

METHODS OF CAUSING EXPLOSIVES TO EXPLODE.

From what has been stated it appears that the different explosives can be caused to explode by various means. All of them can, under some circumstances, be made to explode by fire; some of them can be caused to undergo a detonating explosion if another explosive is detonated in contact with them. In blasting, use is made of both of these means for setting off explosives, the means used being determined by the given conditions. The commercial devices employed for initiating an explosion are known as fuses, detonators, electric igniters, and electric detonators.

FUSE.

The term fuse embraces two different classes of devices, namely, burning fuse and detonating fuse, each being characterized by the material with which it is charged and its consequent behavior in use. Burning fuse as now known was invented in 1831 by William Bickford, of Cornwall, England. It is intended to convey fire to an explosive or combustible mass without danger to the person lighting it. Before its invention, straws and squibs (Pl. V, *A*) containing finely granulated gunpowder were used for igniting charges of blasting powder, metallic needles, similar to those still used in the coal mines of this country, being used to make an opening through the stemming from the charge to the mouth of the drill hole. The straws containing the gunpowder, when pushed into the opening thus made, formed there a powder train by which the charge was ignited. As the rate of burning in this train of powder was fast and irregular, sometimes approaching an explosion, the use of a slow-burning paper to communicate the flame from the ignition lamp to the powder train was found necessary.

ADVANTAGES OF BICKFORD'S BURNING FUSE.

The increased interval that elapsed between the application of the flame from the lamp to the paper and the ignition of the powder train gave the miner more time in which to retreat to a place of safety. Sometimes, however, he was not successful in this, owing to an unusually rapid rate of burning of the paper and the powder train, and therefore he was exposed to considerable risk. Through the invention of the burning fuse by Bickford, the rate of burning was made more regular and certain, the efficiency in firing charges of blasting powder was increased, and the safety of the operation was improved. Consequently, the fuse has been termed "safety fuse," but this term is going out of vogue, because it was used in a purely relative sense, and it gives no assurance of safety under such conditions as exist in gaseous coal mines and other places. It has also been known under the term "running fuse."

KINDS OF BURNING FUSE ON MARKET.

Burning fuse is sent into the market in several varieties, in 50-foot lengths rolled in coils, but each variety consists of a core of mealed gunpowder inclosed in two or more layers of yarn and generally surrounded by tape that has been dipped in a waterproofing composition. (See Pl. V, B.) Some varieties are then dusted with substances like whiting or white clay to prevent the sticky surfaces from sticking to one another. The kinds of fuse known to commerce are hemp fuse, cotton fuse, single-tape fuse, double-tape fuse, triple-tape fuse, single-countered fuse, and double-countered fuse.

MODE OF ACTION OF BURNING FUSE.

When properly made and in perfect condition, if one end of a fuse of any one of the varieties mentioned is lighted, the powder core burns slowly along the fuse until at last the flame runs out at the farther end, and if it touches a charge of powder sets it off.

STANDARD RATE OF BURNING DESIRABLE.

Fuse should be so made and should be in such condition when used that any part of any coil will burn at a rate that does not vary more than 10 per cent above or below the standard rate, the standard rate accepted by the Government in its work being a yard in 90 seconds. A uniform rate of burning such as this is of the greatest importance in blasting, because in setting a charge the blaster or shot firer cuts a piece of fuse which he believes is long enough to reach from the charge to the face and sufficiently far beyond to give him ample time, after the outer end has been set on fire, to reach a place of safety before the flame reaches the charge. In order that a fuse may burn at a regular rate the powder core should be uniform in character and should extend continuously through the fuse.

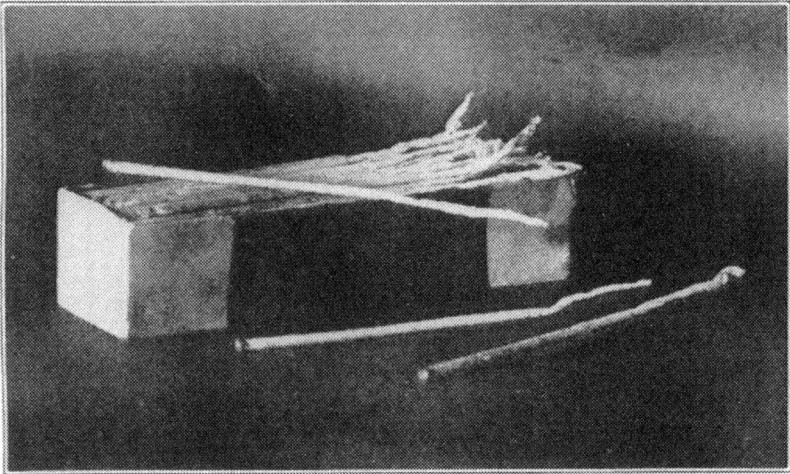

A. MINER'S SQUIB.

B. MINER'S SQUIB AND FUSE.

FUSE SHOULD BE TESTED FOR RATE OF BURNING.

Examination of fuses by the X-ray (Pl. VI, *A*) has shown that sometimes the powder does not extend continuously through the fuse, and that spaces may separate parts of the core; therefore, careful tests should be made of the rate of burning of pieces of any fuse concerning the soundness of which there is any doubt whatever. Although the manufacturers may produce a fuse with a regular rate of burning, the rate may be changed by bad handling, as, for instance, by being squeezed in handling or tamping so as to disturb the powder core; or by being suddenly or roughly opened when the coil is stiff from cold or age, the fuse being thus cracked; or the fuse may be injured by being rubbed against the rough surfaces of the rock.[a] (Pl. VI, *B*). Care should be taken in the storage of fuse, for the rate of burning of a fuse varies not only with the composition of the powder train, its degree of fineness, its hygrometric condition, and the age of the fuse, but also with the conditions under which the fuse has been stored, the manner in which it is used, and the way in which it is tamped into a drill hole. Tests have shown that the rate of burning of fuse stored in a very damp place varied from 17 to 45 per cent; warming the fuse near a stove gave a variation of 30 per cent; and under actual working conditions in rock the rate of burning varied from 22 to 82 per cent, according to the kind of stemming used and the method of tamping employed.

REQUIREMENTS BURNING FUSE SHOULD SATISFY.

The requirements to be met by burning fuse are well outlined in the recent specifications adopted by the United States Government for the purchase of all fuse on the Canal Zone and on the different projects of the Reclamation Service, which are as follows:

All fuse furnished shall be of the type known as "safety fuse," shall be free from defects, and shall be capable of being stored for at least six months without deterioration. It shall be put up in properly labeled packages containing 2 coils each, and the rate of burning in open air (namely, 90 seconds per yard) shall be stated on each wrapper. Sixty-package lots, containing 120 50-foot coils, shall be packed in air-tight wooden cases. At least 118 coils in each case shall be in continuous lengths of 50 feet; the 2 remaining coils may be made up of 2 pieces each. The ends of the two coils last mentioned must be tied together. All safety fuse when burning shall not burst nor explode in any part of its length. It shall burn without any such lateral sparking or glowing at the sides as might cause short-circuiting when the fuse is coiled on itself. When burned in the open air it shall burn quietly and uniformly, the rate of

[a] See Hall, Clarence, and Howell, S. P., Investigations of fuse and miners' squibs: Technical Paper 7, Bureau of Mines, 1912, 19 pp.

burning not varying more than 10 per cent over or under the stated rate (90 seconds per yard). The powder core shall be continuous, without gaps, and of sufficient quantity so that the final spit is strong enough to ignite another piece of fuse when the ends of two pieces are separated at least 1 inch. All safety fuse shall be sufficiently waterproof to stand immersion for not less than 30 minutes in water at least 1 foot in depth. One 50-foot length will be selected at random from each case for the purpose of inspection.

TRANSPORTATION AND STORAGE OF BURNING FUSE.

Burning fuse when transported on railroads is regarded in this country and in foreign countries as ordinary freight. A carload of fuse may contain hundreds of pounds of gunpowder, but as the core of gunpowder is imprisoned between layers of hemp and cotton cord it is deprived, on ignition, of its effective explosive force, and therefore is not a source of danger from explosion. Fuse should be stored in a dry place and in a moderate temperature. It should not be stored in a metal box exposed to the direct rays of the sun or to other sources of heat. The using up of an old lot of fuse before opening a new package insures the best results in practice.

USE OF BURNING FUSE WITH DETONATORS.

Burning fuse is used not only to produce ignitions of blasting powder and similar explosives, the fuse being burned in direct contact with charges of explosives, but also to effect detonation in charges of high explosives. For the latter use the end that is to be laid toward the charge in the bore hole is inserted within a detonator, and for this reason the diameter of the fuse is made such that the fuse will neatly fit into the capsule containing the fulminate or other detonating charge. In selecting burning fuse for use it must be remembered that the moisture conditions under which blasting is done vary widely; they range from the blasting of very dry material to submarine work, in which the material to be blasted is completely under water. Hence manufacturers of fuse endeavor to make varieties adapted to the various conditions.

CLASSES OF BURNING FUSE.

In this country five classes of burning fuse are made, as follows:
1. Fuse for use in dry material.
2. Fuse for use in damp material.
3. Fuse for use in wet material.
4. Fuse for use in very wet material.
5. Fuse for use in submarine work.

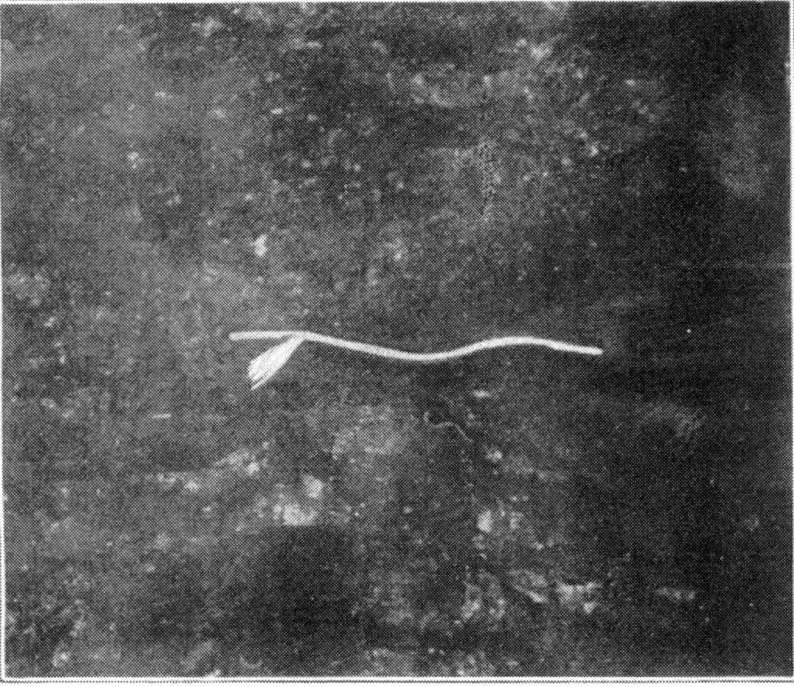

B. SIDE SPITTING OF A FUSE.

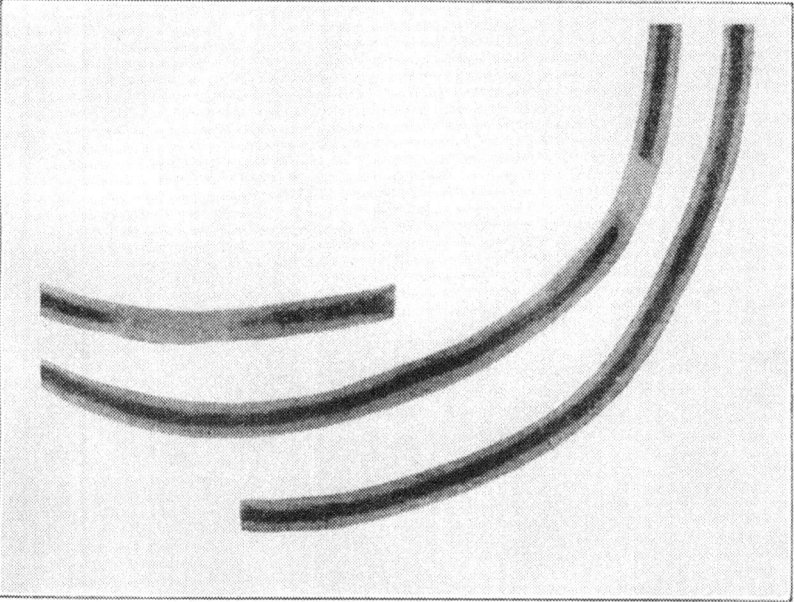

A. X-RAY PRINT OF DEFECTIVE FUSE.

Class 1 is cheaply constructed, has no waterproof qualities, and requires great care in manipulation. Its use is small and it is not considered reliable.

Class 2 is better than class 1, but has not all the requirements of a good fuse. It will stand immersion under a few inches of water for a few minutes, but it is not recommended for general use. This class, as well as class 1, is invariably subject to lateral sparking when burning.

Classes 3, 4, and 5 are well made. They will stand immersion in shallow water 30 minutes, 24 hours, and 96 hours, respectively. In burning they show little, if any, lateral sparking or glowing. They usually burn quietly and steadily and will not burst nor explode. In metal mines, where it is often necessary to use different lengths of fuse at the same time in order to fire dependent shots, the projecting ends of the different lengths, if the fuse is of these grades, can be coiled and placed in the mouth of a drill hole to keep them from being cut by flying fragments of rock. The lateral sparking is rarely sufficient to ignite adjacent fuse or cause short-circuiting, and accordingly the practice can be followed with comparative safety.

MARKING DIFFERENT FUSE.

To meet the varying conditions that arise in practice, manufacturers produce grades of fuse having different rates of burning, though each grade may have a uniform rate of burning. Tests of the various fuses sold in this country have shown that the rates of burning in the open air for the different grades range from 54 to 120 seconds per yard. In some makes this difference is distinguished by the color of the paper wrapper in which the fuse is enveloped, but in many makes there is no indication on the wrapper or fuse of the rate of burning, and without such information a miner who has been accustomed to use a certain brand of fuse may inadvertently use another that burns faster, with the result that a charge may be exploded prematurely, and all connected with the work may be endangered. Whenever there is any doubt whatever as to the rate at which a fuse will burn, before it is used in blasting, a section at least a yard long should be placed on a board on the ground in the open air and burned, and the rate of its burning from the time when it is ignited at one end until the spit occurs at the other end should be determined with a watch.

CARE TO BE USED IN CUTTING FUSE.

Fuse should be carefully handled and should never be laid on a damp place before being used. In cutting fuse and in fitting it in place care must be taken that the powder core does not run out for that might cause a misfire. If a fuse is to be inserted in a detonator, the fuse should be held upright and cut straight across with a sharp knife before the end is placed in the detonator. It should never be forgotten that with all the care exercised by the manufacturers of " safety fuse " to perfect the quality of fuse, human life depends in a great measure on the proper use of fuse by the miner himself. He should have a better understanding than he usually has of the action of fuse, because under certain conditions there is always a risk in its use.

DETONATING FUSE.

Although it has been shown that, by being attached to a detonator, burning fuse may be employed in effecting detonation, and although it is largely so employed in blasting and mining, there has been devised a variety of fuses that undergo detonation throughout their length when once detonation has been started in them. Such fuses have been made by filling a lead tube with guncotton or with compositions containing mercury fulminate and, by the aid of a draw plate, drawing this tube, thus charged, down to the desired diameter. Although such fuses have been used to some extent in military-engineering operations, they have met with little success in ordinary practice.

CORDEAU DETONANT.

In recent times lead tubes charged with trinitrotoluene have been drawn out into fuse in the manner described. Such detonating fuse is known as cordeau detonant and cordeau Bickford and has come to be used to a considerable extent for testing the rate of detonation of high explosives, and to some extent has been employed in deep holes, where sections of the fuse have been laid beside the charges of high explosive introduced into the holes. The object of its use is to accelerate the rate of detonation of the charge so as to enable it to break the inclosing rock more effectively. In firing large blasts cordeau detonant has been used to connect the charges in the various bore holes, so that by the detonation of the central charge detonation would be communicated through the fuse to the surrounding charges. As many as 2125 holes have been shot at once in a single blast by its use.[a] Up to the present its use in industrial practice has been limited, and the use of detonating fuse of this most successful variety in engineering practice may be held to be still only an

[a] Barthélemy, L., Nouvelle application du cordeau détonant aux travaux publics : Bull. Soc. eng. civ. France, November, 1910, p. 492.

emergency device for use under special conditions. The characteristics of cordeau detonant are as follows:

Characteristics of cordeau detonant.

Outside diameter of tube_____inches__	0. 1645
Thickness of lead_____do____	. 0250
Inside diameter of tube_____do____	. 1145
Weight of a foot length of fuse_____grams__	25. 77
Weight of a foot length of lead tube_____do____	23. 09
Weight of a foot length of charge_____do____	2. 68
Density of charge in fuse_____	1. 32

Consistency of charge: Powdered, very fine, dry, soft, slightly cohesive.
Color of charge, straw.

BOOSTERS.

Recently brass tubes about 4 inches long, containing a charge of trinitrotoluene, and of such internal diameter that the fused detonator may slip neatly into the tube, have been put upon the market under the name of renforts or boosters. A booster is intended for insertion in the primer cartridge of a blasting charge, especially when gelatin dynamites containing a high percentage of nitroglycerin are used, as the booster increases the speed of detonation of such charges.

CORDEAU BICKFORD.

Cordeau Bickford is another type of detonating fuse. It is sent into the market wound on spools containing 100 to 500 feet each, the exact length being specified on the spool head. The spools are provided with a hole in the center through which a rod may be run for convenience in unwinding. They are packed for shipment in neat, strong, wooden boxes. In France, England, and the United States this fuse is accepted for shipment by the transportation companies without restrictions, except that it shall not be packed with detonators or high explosives. Cordeau Bickford of standard American size (0.198 inch diameter) weighs about 7 pounds per 100 feet. Cordeau Bickford is accompanied with certain accessories for use in making up the connections, such as cordeau slitters for splitting the end of branch lines, cordeau crimpers for cap crimping and fuse cutting, straight unions for connecting an ordinary cap to the cordeau Bickford, special unions for connecting electric detonators to the cordeau Bickford, couplings for connecting two lengths of this fuse, and sleeves for special connections.

DETONATORS.

Detonators, which are also called blasting caps, or sometimes, in mining coal, exploders, consist of copper capsules of about the diameter of an ordinary lead pencil that are commonly charged with dry mercury fulminate or with a mixture of dry mercury fulminate and potassium chlorate, the charge being so compressed in the bottom of the capsule as to fill it to about one-third its length. In recent times other detonating substances have been substituted to a greater or less extent for the mercury fulminate in these devices. Among the substances used are trinitrotoluene (called trotyl), tetranitromethylaniline (called tetryl), and the metallic hydronitrides, especially the lead compound, PbN_6, and detonators containing compositions of this kind have made their appearance in the market.

ADVANTAGES OF USING DETONATORS IN FIRING HIGH EXPLOSIVES.

Dynamite and other detonating explosives are in practice fired by means of detonators, for though they may be exploded by fuse or gunpowder primers, yet the explosion so produced is not complete; the explosives are not then used to the best advantage, all the work they are capable of doing is not done, and, moreover, the gases that are produced from them under these conditions are usually dangerous, owing to poisonous qualities. When high explosives are employed it is not only safer to fire them with detonators strong enough to insure their own complete explosion, but it is surer and cheaper, as the explosives do more and better work.

GRADES OF DETONATORS.

Several grades of detonators, also called blasting caps or exploders, are to be found in the market. The "strengths" of detonators most commonly used, measured by the weights of fulminating composition contained in them, are as follows:

Grades of detonators and weights of their charges.

Grade of detonator.		Weight of charge.	
Testing-station.	Commercial.	Grams.	Grains.
No. 3...	3X, or triple......	0.54	8.3
No. 4...	4X, or quadruple..	.65	10.0
No. 5...	5X, or quintuple..	.80	12.3
No. 6...	6X, or sextuple....	1.00	15.4
No. 7...	7X, or No. 20......	1.50	23.1
No. 8...	8X, or No. 30......	2.00	30.9

DETONATORS WEAKER THAN NO. 6 SHOULD NOT BE USED.

Under date of July 15, 1914, the director of the bureau was informed by the secretary of the Institute of Makers of Explosives that at the last meeting of the institute the resolution of the committee recommending the elimination of the manufacture and sale of caps weaker than No. 6 strength was unanimously adopted, and that the secretary had been instructed to write a letter to the manufacturers of detonators informing them of the action of the institute.

METHOD OF ATTACHING BURNING FUSE TO DETONATORS.

Detonators are fired by the aid of a piece of burning fuse. In practice a piece of fuse of the desired length while held upright is cut straight across with a sharp knife just before the end is placed in the detonator. In a humid atmosphere at least 1 inch of the exposed end of the fuse coil is cut off and thrown away before the required length is inserted in the detonator, care being taken in cutting the fuse that the powder core does not run out of the fuse. The fuse is held vertically upward, the upper end is gently inserted in the mouth of the blasting cap and carefully pushed in until it just touches the surface of the detonating composition, and is then crimped tightly in place in the capsule. (Pl. VII, A.) Because detonators are used so largely in connection with fuses in firing charges in quarries, tunnels, and mines it is customary to make detonators and fuses in standard sizes so that the fuse may easily slip within the mouth of the detonator and yet make a close fit within it. In use with explosives the detonator with its attached fuse is inserted in the charge to be fired and is well secured to it. The whole is then placed in the bore hole, covered with stemming, and tamped. It is then ready to be fired. In firing, the end of the fuse is ignited and the fire rushes down the powder core until it streams against the detonating composition, which then ignites and detonates and causes the detonation of the explosive with which it is in contact.

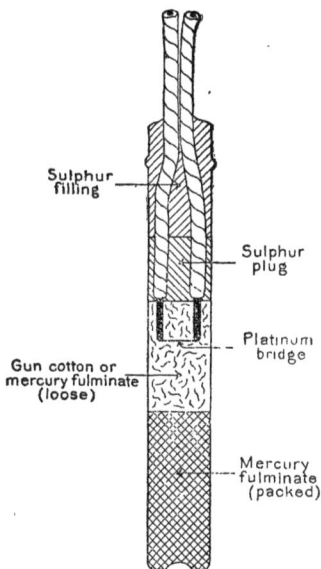

FIGURE 1.—Electric detonator, showing its component parts.

ELECTRIC DETONATORS.

Electric detonators (fig. 1) are simply ordinary detonators that have been fitted with a means of firing them by the electric current. This is done by inserting within them two copper wires joined at the inner ends by an extremely fine platinum or other high-resistance wire, which, like the carbon filament in the ordinary incandescent lamp, becomes heated till it glows, when an electric current is passed through it. This wire, known as the bridge, is placed above the detonating composition and is surrounded by guncotton or loose fulminate. The space above it and the mouth of the capsule are then filled and closed by means of a plug of sulphur or other water-proof composition, which is poured in while soft. The copper wires pass through the plug and are long enough to extend outside the capsule. These outer ends are called the legs or wires of the electric detonator. Although the copper wires are bare within the electric detonator, the legs outside are covered with an insulating wrapping. These legs are made of different lengths in order to suit different depths of bore holes. The charge of detonating composition differs in the different grades of electric detonators so as to give different strengths.

GRADES OF ELECTRIC DETONATORS.

The following table gives the grade and weight of charge for the more common electric detonators:

Grades and weights of charge of electric detonators.

Grade of detonator.		Weight of charge.	
Testing-station.	Commercial.	Grams.	Grains.
No. 5.....	Single strength........	0.80	12.3
No. 6.....	Double strength.......	1.00	15.4
No. 7.....	Triple strength........	1.50	23.1
No. 8.....	Quadruple strength...	2.00	30.0

FIRING ELECTRIC DETONATORS.

In loading bore holes electric detonators are placed in the charge just as detonators with fuse are, and the bore holes are tamped in a similar manner. To fire the charge, the legs of the detonator are connected by leading wires to an electric device at a safe distance, and from it the current is sent to fire the blast. No flame can escape from the bore hole during the firing if the stemming fills the hole completely, and hence blasting in gaseous mines is made much safer and all blasting made more certain and efficient.

When the priming cartridge, with its fused detonator or electric detonator inserted in it, is being pushed into the bore hole, or later when the stemming is being loaded and tamped about it, unless the detonator is firmly secured in the priming cartridge, there is danger that the detonator may be pulled partly or wholly out of the cartridge, and it is found that if there be even a slight separation be-

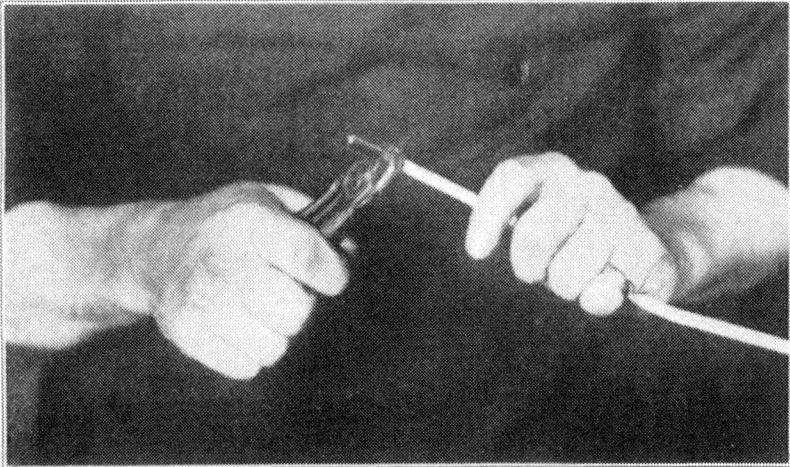

A. CRIMPING DETONATOR ON FUSE.

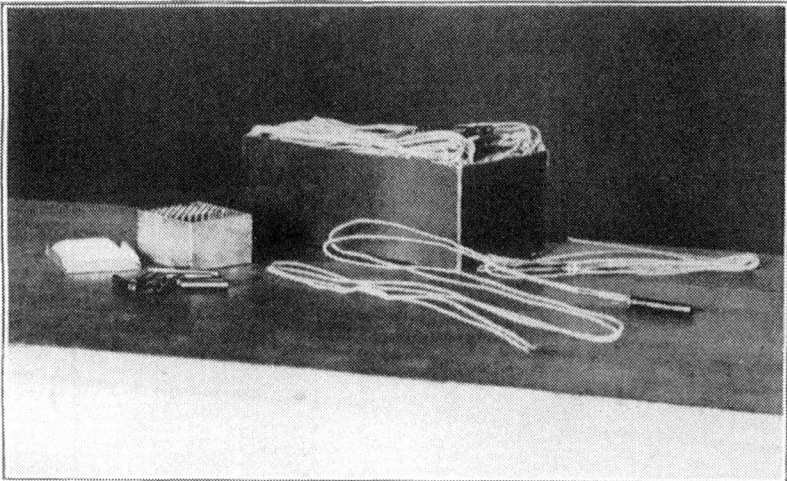

B. DETONATORS AND ELECTRIC DETONATORS.

tween them, an imperfect detonation, or a burning of the charge, or even a misfire, may result. Where electric detonators or igniters are made use of, the system may be made fast by taking a bight around the cartridge with the legs of the detonator or igniter, but this exposes the legs to abrasion, so that at times their insulation may be rubbed off.

CLIP FOR DETONATORS.

For use with fused detonators or electric detonators W. Cullen[a] has devised a clip (Pl. VIII, A and B) consisting of a circular sheet of thin malleable metal with three equidistant arms projecting from the circumference and a star-shaped pronged hole punched in its center. In use the fuse is pushed through the hole, the detonator crimped on by pliers, the whole inserted in the primer through a hole punched through the end of the wrapper, and then the arms are folded down about the cartridge. The prongs prevent the fuse and detonator from being pulled through the star-shaped hole. The clip is for use only in quarries and metal mines.

DELAY-ACTION ELECTRIC DETONATORS.

What are called delay-action electric " exploders," or detonators, are now being offered for use where a number of holes are to be fired in one operation, but in such a way that the charges will explode one after another. The exploder consists of an electric igniter attached to a short length of powder fuse penetrating a detonator, the whole being inclosed in a waterproof covering. Periods of 10 different lengths are provided in exploders, the different periods being obtained by varying the length of the fuse that is inclosed. The longest of these delay-action electric detonators is short enough to be inclosed in a stick of explosive. They are manufactured with any desired length of legs.

DANGER IN USE OF DETONATORS.

In the description of mercury fulminate attention was called to its extreme sensitiveness to heat, friction, or blows, and to the extreme violence of its explosion. All these properties therefore belong to detonators and electric detonators, and these devices should be treated with the utmost respect. Never attempt to pick out any of the composition. Do not drop them or strike them violently against any hard body. Do not lay them on the ground where they may be stepped on. Do not step on them. In crimping, take the greatest

---

[a] Cullen, William, Discussion of the Cullen detonator clip: Jour. Chem., Met. and Min. Soc. South Africa, vol. 14, February, 1914, pp. 366–367.

care not to squeeze the composition, and never crimp with the teeth, for there is enough composition in one of these small capsules to blow a man's head open.

STORAGE AND TRANSPORTING OF DETONATORS.

Detonators should be stored in a dry place and in a building apart from any other explosives. They should never be carried into a mine with other explosives, and they should never be placed in a mine near other explosives except in bore holes. When carried or shipped, they should be packed firmly with a quantity of elastic material, such as felt or the coiled legs of the electric detonators, about them, and they should not be exposed to heat, blows, or shocks of any kind. Plate IX, A, shows a new type of container for detonators. The device prevents the detonators from coming in contact with one another or with the metal of the container.

ADVANTAGE OF CONFINING MERCURY FULMINATE.

It has been observed that mercury fulminate in small amounts, when ignited, develops its full force only when confined. It has been believed also that the sulphur plugs used in electric detonators more completely confine the fulminating composition of such detonators than do pieces of fuse inserted in the detonators, even when the fuses are properly used and securely crimped in place. It seemed desirable to demonstrate the accuracy of these views by making comparative efficiency tests of electric detonators closed by sulphur plugs and of detonators fitted with fuse, and for this reason they were tested by the "nail test,"[a] which was chosen for the reason that it appears to approximate more nearly the results established for the efficiency of detonators in practice than does any of the several other direct methods devised.

NAIL TEST FOR DETONATORS.

In the nail test finishing nails, 4 inches long, are used as indicators of the efficiency of the electric detonators. For the tests here reported, the nails were so selected that they were of approximately the same length, the same gage, and the same weight. The bottom of the detonator was placed 1¾ inches from the face of the head of the nail and was laid parallel to the nail, being separated from it by two 22-gage (0.025-inch) copper wires that were wrapped

---

[a] Hall, Clarence, and Howell, S. P., Investigations of detonators and electric detonators: Bull. 59, Bureau of Mines, 1913, pp. 25, 26.

A. CLIP FOR DETONATOR. 1, CLIP; 2, DETONATOR; 3, FUSE; 4, FUSE WITH CLIP IN PLACE AND DETONATOR CRIMPED ON; 5, PRIMER WITH DETONATOR INSERTED IN IT AND CLIP CRIMPED ON.

B. CLIP FOR ELECTRIC DETONATOR. 1, CLIP; 2, DETONATOR; 3, LEAD WIRES WITH CLIP IN PLACE AND DETONATOR ATTACHED; 4, PRIMER WITH DETONATOR INSERTED IN IT AND CLIP CRIMPED ON.

A. NEW TYPE OF DETONATOR BOX. COVER REMOVED TO SHOW METHOD OF PACKING DETONATORS IN ORDER TO PREVENT THEIR COMING IN CONTACT WITH ONE ANOTHER OR WITH METAL PART OF BOX.

B. FACE OF ANCON QUARRY.

around the detonator. The detonator was fastened in position by one strand of a similar copper wire, which was wrapped around the detonator and the nail (see fig. 2), midway between the ends of the detonator. The whole was suspended horizontally in the air in such a manner that the nail was directly above the detonator, which was then fired. The effect of the explosion of the detonator in contact

with the nail was to bend the nail and to project it upward. Care was taken that the nail was not hurled against any solid surface in such a manner as to be further bent.

DETAILS OF SPECIFIC NAIL
TESTS.

Five trials were made with each electric detonator. The angle through which the nail was bent from its normal position around its major axis was

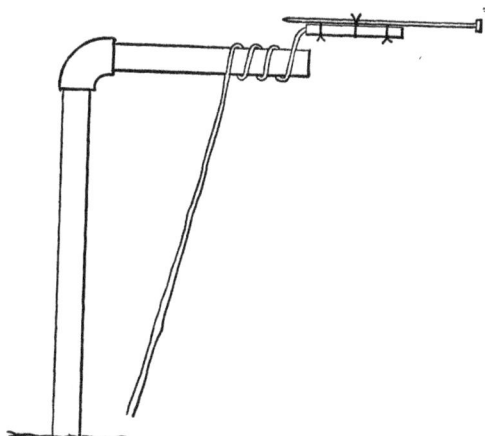

FIGURE 2.—Nail in position for test of electric detonator.

measured, and the average angle in five trials was taken as a measure of the efficiency of the detonator. Nails used in the tests are shown in Plate X.

The nail test was made with No. 3 and No. 6 detonators and with electric detonators constructed from detonators taken from the same lot as these, and the following results were obtained:

Results of nail tests with No. 3 and No. 6 detonators and with electric detonators.

Grade of detonator or electric detonator.	Test No.	Angle of bending resulting from trial No.—					Average.
		1	2	3	4	5	
	M—	°	°	°	°	°	°
No. 3 detonator a	289	8	7	9	9	8	8.2
No. 3 electric detonator	279	12	10	8	9	7	9.2
No. 6 detonator a	318	32	29	35	30	31	31.4
No. 6 electric detonator	321	31	31	33	36	27	31.6

a Fired with fuse placed against the compressed charge of mercury fulminate composition and crimped in place.

These tests show that with low-grade detonators, containing only small charges of fulminating composition, the efficiency of electric detonators is slightly greater than that of " fused " detonators, but with higher grade detonators, containing larger charges, such as the No. 6, the efficiency of both varieties is practically the same.

This indicates that the additional confinement given by the plug of the electric detonator over that afforded by the fuse of the "fused" detonator is important only with the low-grade detonators.

FAILURES FROM USE OF LOW-GRADE DETONATORS.

The investigations of detonators and electric detonators by the Bureau of Mines have shown that the average percentage of failures of explosives to detonate was increased over 20 per cent when the lower grades of electric detonators were used in place of No. 6 electric detonators, and over 50 per cent when they were used in place of No. 8 electric detonators.

AGED AND INSENSITIVE EXPLOSIVES RENDERED EFFICIENT BY USE OF HIGH-GRADE DETONATORS.

In a number of tests with aged explosives the efficiency was also found to increase with the grade of detonator used. For example, the average efficiency of four explosives of different ages was increased 10.4 per cent when a No. 6 detonator was used with them in place of a No. 4 detonator. It is noteworthy, however, that when sensitive explosives were tested under conditions favorable to detonation, practically the same energy was developed, irrespective of the detonator used. These tests emphasize the importance of using explosives in a fresh condition, but, as this is not always possible in practical work, in order to offset the effect of deterioration of explosives due to aging, strong detonators should be used.

These observations also indicate that a less sensitive, and, consequently, safer and probably cheaper, explosive may be employed if a sufficiently high grade of detonator be used to start its explosion.

These observations have also a bearing on the use of detonators in other ways, especially when distributed through a charge.

TESTS WITH DETONATORS DISTRIBUTED IN CHARGE.

In some sections where long charges of high explosives have been used in blasting, it has been the custom to place two or more separated detonators in the charge in the belief that the work accomplished by the explosive was thereby increased. To test the value of this practice "rate-of-detonation tests" were made with an explosive of class 1, subclass a^a, put up in cartridges $1\frac{1}{4}$ inches in diameter, with and without No. 7 electric detonators distributed in the explosive, in order to determine whether the presence of the detonators within the charge would increase the distance through which detonation of the charge was effected. The explosive used was subsequently tested in cartridges that were $1\frac{1}{2}$ inches in diameter, with

---

[a] Part of classification by Bureau of Mines. It signifies an ammonium nitrate explosive containing a sensitizer that is itself an explosive.

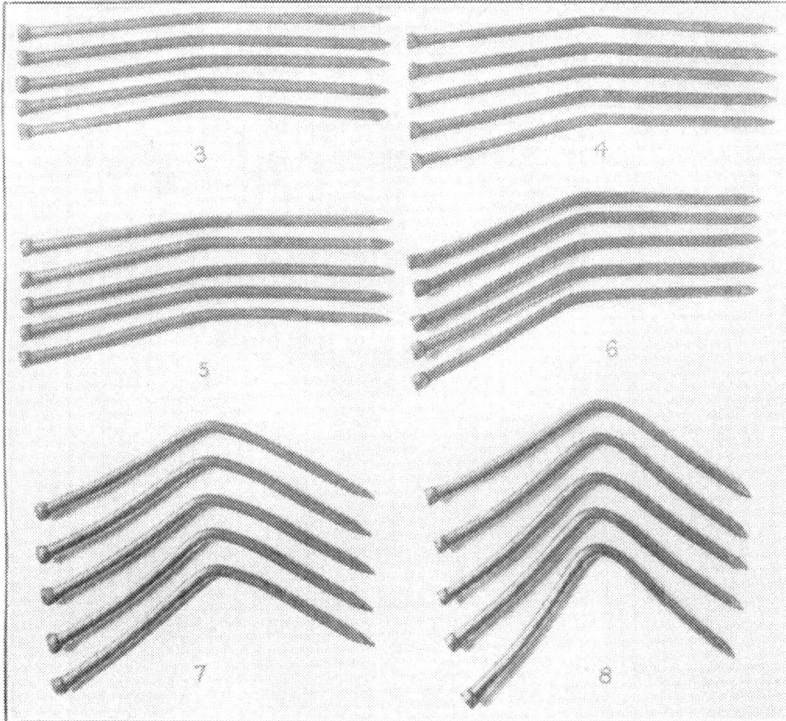

A. RESULTS OF NAIL TESTS OF P. T. S. S. ELECTRIC DETONATORS NOS. 3, 4, 5, 6, 7, AND 8.

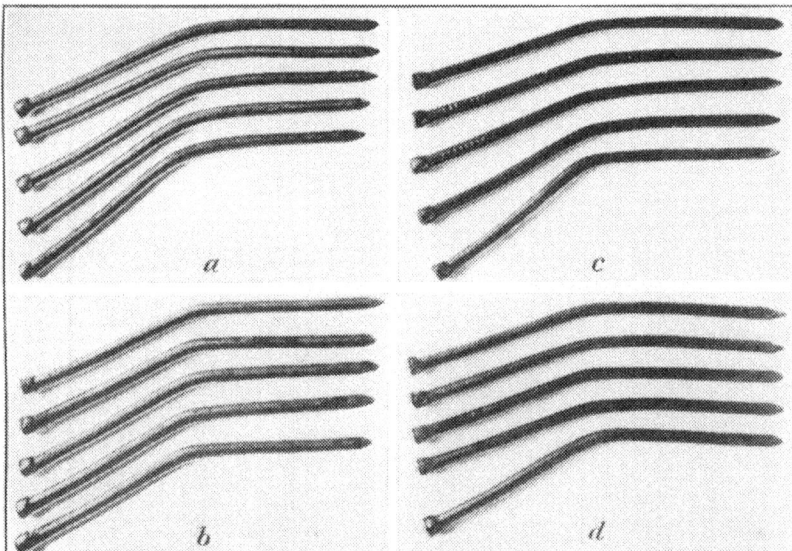

B. RESULTS OF NAIL TESTS OF NO. 6 ELECTRIC DETONATORS. *a*, WESTERN COAST; *b*, SPECIAL; *c*, P. T. S. S.; *d*, FOREIGN.

the result that the No. 7 electric detonator caused complete detonation in every trial, whereas the No. 3 detonator failed to do so once out of three trials. The explosive named was purposely chosen, because it was relatively insensitive to detonation. The results of the tests follow.

Results of rate-of-detonation tests in which No. 7 electric detonators were distributed in the charge.

Test No.	Result.	Dimensions of galvanized-iron tube used.	
		Diameter.	Length.
		Inches.	*Inches.*
D1134 [a]	Detonation incomplete	$1\frac{1}{2}$	42
D1147 [b]	.....do.	$1\frac{3}{4}$	80
D1148 [c]	.....do.	$1\frac{1}{2}$	42

[a] No electric detonators distributed in the charge.
[b] Three No. 7 electric detonators placed every half meter in the charge.
[c] Three No. 7 electric detonators placed every quarter meter in the charge.

Further tests were made with an insensitive gelatin dynamite, by placing seven No. 7 electric detonators every eighth meter in the charge. The results were as follows:

Results of rate-of-detonation tests in which No. 7 electric detonators were distributed in the charge.

Length of charge that detonated.	Diameter of cartridges.	Dimensions of galvanized-iron tube used.	
		Diameter.	Length.
Inches.	*Inches.*	*Inches.*	*Inches*
$21\frac{1}{2}$	$1\frac{1}{2}$	$1\frac{3}{4}$	42
16	$1\frac{1}{2}$	$1\frac{3}{4}$	42
20	$1\frac{1}{2}$	$1\frac{3}{4}$	42

Tests with the same explosive when no extra electric detonators were used gave the following results:

Results of rate-of-detonation tests in which no extra electric detonators were used.[a]

Length of charge that detonated.	Diameter of cartridges.	Dimensions of galvanized-iron tube used.	
		Diameter.	Length.
Inches.	*Inches.*	*Inches.*	*Inches.*
6	$1\frac{1}{2}$	$1\frac{3}{4}$	42
15	$1\frac{1}{2}$	$1\frac{3}{4}$	42
9	$1\frac{1}{2}$	$1\frac{3}{4}$	42

[a] No. 7 electric detonators used for firing the charge.

These results indicate that when extra electric detonators are distributed 5 inches apart in a 40-inch file of cartridges of an insensitive explosive they have a slight tendency to increase the rate of propagation of the explosion wave, but when the extra electric detonators are placed 10 inches apart they offer no advantage.

With this explosive the effect of the embedded electric detonator seemingly did not extend farther than 5 inches. Probably the explosion wave, induced by the primary electric detonator fired, caused the detonation of the explosive in advance of that of the included electric detonator. If this be true, the included electric detonator is exploded within the products of explosion of the explosive, and therefore offers little, if any, aid in starting the detonation of any part of the charge.

SIMULTANEOUS FIRING OF SEVERAL ELECTRIC DETONATORS IN A CHARGE.

Attention is called to the fact that it is sometimes advantageous to fire at the same time two or more electric detonators placed in different parts of the charge in the same drill hole. If one electric detonator is placed in the bottom of the charge and one in the top, and they are fired at the same time, the time of detonation of the charge of explosive is reduced approximately one-half, and, accordingly, the shattering effect of the explosion is materially increased. When long charges of explosives are used it is sometimes necessary to use more than one electric detonator in the charge to insure complete detonation.

TWO KINDS OF EXPLOSIVES IN THE SAME DRILL HOLE.

In certain quarry operations in the Middle West, owing to the variations in the hardness and structure of the different strata, it is advantageous to use more than one kind of explosive in the same drill hole. The part of the drill hole that penetrates the hardest stratum is usually loaded with an explosive having a high rate of detonation. The remainder of the charge may be an explosive having an intermediate rate, or a black blasting powder may be used, if there are no pronounced clay seams or other irregularities that would allow the gases evolved on the explosion of the black blasting powder to escape before the main charge detonated. In work of this kind the holes are drilled vertically 15 to 20 feet deep, and there is always sufficient stemming used to insure the maximum effect of the blast, even when the explosives used in the same drill hole detonate at different rates.

DANGER FROM USE OF COMBINATION CHARGES IN COAL MINES.

This practice of using combination charges of explosives has been recently adopted in some of the coal mines of the anthracite field of Pennsylvania, but most of the drill holes are shallow, and hence do not permit the use of sufficient stemming to properly confine the gases produced by the explosion when they are evolved at different rates. When explosives are used under these conditions fires and blown-out shots are likely to result.

NO ADVANTAGE FROM COMBINATION CHARGES IN COAL MINES.

Several tests made at the Pittsburgh experiment station of the bureau to determine the energy developed by combination charges showed that there is no advantage in using them for blasting in coal mines. In some of the tests a No. 6 detonator was inserted in the charge of dynamite and the cartridge so placed that the detonator was in the back of the bore hole. In front of this a charge of black blasting powder containing a black-powder igniter was placed, and the empty part of the drill hole was then well tamped with clay stemming.

BALLISTIC-PENDULUM TESTS.

Tests of combination charges were also made with the ballistic pendulum.[a] The ballistic pendulum (Pl. XI, B) is used to measure the deflective force of explosives. The apparatus consists of a 12.2-inch mortar suspended as a pendulum from a beam having knife edges, supported by two massive concrete piers, a cannon mounted on a truck, a track upon which the truck rests, and an automatic recording device. The mortar weighs 31,600 pounds and hangs in two **U**-shaped rods. The rods pass through and are held by two cast-steel saddles, which fit over the supported beam. The radius of the swing measured from the point of the knife edges to the center of the trunnions is $89\frac{3}{4}$ inches.

The cannon is 24 inches in diameter and 36 inches long, and its axial bore is $2\frac{1}{4}$ inches in diameter and $21\frac{1}{2}$ inches deep.

The swing of the mortar is measured by an automatic recording device which consists of a movable scale and a vernier set on a metal base fastened to a concrete footing. The scale is actuated by a long rod set in guides which bears against a stud bolt in the bottom of the mortar directly below its center of gravity. The scale moves tangentially to the swing of the mortar; the radius of the swing, measured from the knife-edge bearings to the center of the rod or base of the stud, is $114\frac{7}{16}$ inches. The swing is measured to one-hundredths of an inch.

---

[a] See Hall, Clarence, and Howell, S. P., The selection of explosives used in engineering and mining operations: Bull. 48, Bureau of Mines, 1913, pp. 45–46.

The explosive to be tested is charged in the cannon without air space around the charge. One pound of dry fire clay is used as stemming to confine the charge.

RESULTS OF TESTS.

Following is a record of results obtained when the tests were made in the ballistic pendulum with combination charges of 40 per cent "straight" nitroglycerin dynamite and FFF black blasting powder with and without a No. 6 detonator embedded in the dynamite.

The swing of the ballistic pendulum in tests in which a detonator was used was 3.42, 3.41, 3.40, 3.41, 3.26, 3.32, 3.01, 3.34, and 3.28 inches, the average being 3.32 inches. In tests in which no detonator was used the swing was 3.58, 3.30, 3.32, 3.38, 3.24, 3.31, 3.22, 3.36, and 3.31 inches; average, 3.34 inches.

NO ADVANTAGE FROM EXTRA DETONATOR IN DYNAMITE.

These results indicate that there is no advantage in using an extra detonator in the dynamite. Moreover, this practice is also dangerous, for many accidents have occurred in coal mines while combination charges containing detonators were being loaded. This danger increases when squibs are used for firing, for it is then necessary to insert a needle through the stemming and into the charge of black blasting powder, and there is always a possibility of the needle coming in contact with any detonator that is present in the charge.

The practice of using combination charges in coal mines offers no advantage, and, as there are many dangers attendant upon their use, the practice should be discouraged.

ELECTRIC IGNITERS.

Electric igniters are similar in construction to electric detonators except that the capsules are made of paper or wood and are charged with a gunpowder mixture instead of a fulminate.

A. SECONDARY BATTERY, FIRING MACHINE, AND DRY CELL.

B. BALLISTIC PENDULUM.

FIRING BLASTS BY ELECTRICITY.

DANGER OF FIRING BY FLAME.

The proper methods to be used in causing the explosion of charges in blasting depend on the nature of the explosive. To cause the explosion of an explosive of the black-powder class it is necessary only for a flame to touch the explosive, but to cause the detonation of high explosives so that they will have their greatest breaking effect a violent shock is necessary. Black-powder charges are set off by means of squibs, fuse, or electric igniters. Squibs and fuse are set on fire by means of the flame of the miner's lamp, or sometimes by heating a wire to a bright-red heat in the miner's lamp and applying it to the match of the squib or to the cut end of the burning fuse; but these methods would clearly be dangerous if applied in gaseous mines, and they should never be used in any gaseous or dusty mines.

CONNECTING LEGS TO LEADING WIRES AND MAKING SPLICES.

An electric igniter or an electric detonator should be so loaded into a bore hole that when it is in perfect contact with the charge the legs of the igniter or detonator used extend at least 6 inches beyond the completely stemmed and tamped hole. Both legs should be bared of their insulation for about 2 inches from their ends and the wires so cleanly scraped that a good electrical contact can be made with them. Each leg should then be firmly connected with one of the leading wires by about five turns. It is bad practice to have two splices directly opposite each other, because when the leading wires are pulled the splices may touch one another and thus make a short-circuit, which will prevent the electric igniter or electric detonator from being exploded, hence the splices should be stepped apart. A better plan is to wrap the bare-wire splices with tape made for the purpose, which will completely insulate them.

CONNECTING LEADING WIRES TO FIRING MACHINE.

After the legs have been spliced to the leading wires the wires are connected to the firing machine, from which the electric current is to be obtained. This last connection should never be made until all the men are at a safe distance from the place where the blast is to be fired. The rule should be made and never broken that when bore

47

holes are charged the "connecting up" shall move from the bore hole back to the firing machine. The work in all blasting operations should be so organized that it can never be possible for the leading wires to be coupled to the firing machine while anyone is about the place where the holes are being charged and where the blast is to be fired.

CONNECTING DETONATORS IN SERIES.

In heavy blasts in development work two or more electric detonators may be used to good advantage in the same bore hole. For this purpose they are connected in series, which means, for two detonators, that a leg of each is bared and the two bared legs twisted together and wrapped with insulating tape, and that then the two free legs are attached to the leading wires, just as is done when only a single detonator is used. In coal mining the charge used should never be so large as to require the use of more than one detonator in the same hole.

DETONATORS (BLASTING CAPS) DISTRIBUTED IN CHARGE.

When long charges of high explosives are used in blasting operations it has been the custom some times to place detonators (blasting caps) at intervals in a charge, with the belief that the work accomplished by the explosive would be thereby increased. As previously stated, the results of tests made at the Pittsburgh experiment station of the bureau show that this assumption is incorrect and that no advantage results from the use of such detonators in the charge. The direct effect of the first detonator fired at the top of a charge extends only a short distance from its position in the charge, but the explosion wave set up in the explosive by the explosion of the detonator usually proceeds through the charge. When extra detonators are used they are exploded after the explosion of the charge surrounding them and therefore produce little, if any, increase in the efficiency of the blast. However, two or more electric detonators can be used to advantage when they are fired at the same time in charges of explosives more than 5 feet in length, for they may insure the complete detonation of the entire charge and, when distributed through the charge, may tend to reduce the time of the explosive reaction, and thus they materially increase the shattering effect of the explosive.

CONNECTING HOLES.

When it is desired to fire two or more holes at the same time the electric detonators for these holes should be connected in series. To bridge the space between the holes a cheap insulated wire, known as connecting wire, which is not as heavy as the leading wires, may be used. In connecting a series of holes for this purpose, one leg of an electric detonator is connected to one leg of the electric detonator in the next hole and so on to the last hole. There is then left one free leg in the first hole and one free leg in the last hole, and these are spliced to the leading wires.

DANGER FROM A BREAK OR SHORT-CIRCUIT.

The usual precautions should be taken to wrap all the splices with insulating tape so as to completely insulate them and thereby insure a good circuit for the current. It should be borne in mind that the greater the number of holes to be fired in a single blast the greater is the necessity for making sure that the circuit is complete throughout, because if there is a break or a short-circuit at any point, the blast may fail to fire. The delay, expense, and danger caused by such a failure can be prevented by giving from the outset careful attention to the charging and the wiring.

SOURCES OF CURRENT.

The electric current for use in firing electric igniters or electric detonators may be obtained either from primary batteries, such as dry-cell batteries, or from secondary batteries, such as storage batteries, or from electric-lighting circuits, or from generators known as electric firing machines. (Pl. XI, A.)

DRY CELLS.

Firing charges of explosives by means of ordinary dry cells has been prohibited in foreign countries because premature firing of detonators, and sometimes of the charge, has been caused by the wires coming into contact with the poles of the batteries. Safety-contact dry-cell batteries have lately been introduced. These are made with a spring-key contact or with two safety-spring contact buttons, which are the poles of the battery. The two leading wires are laid on the buttons, which are at the same time pushed downward. When the pressure of the thumbs is released the contact is broken. If the wires of a detonator accidentally come into contact with the poles of the battery, the current can not be discharged unless both poles are pushed downward at the same time.

SMALL FIRING DEVICES ARE PORTABLE.

Dry cells, small batteries, and some firing machines are sufficiently light to be carried about by the miner who is to act as shot firer, and this feature insures him against premature firing by any other person.

However, such small devices can at best be used when there are only a few shots in one circuit. The number of shots to be fired and the length of leading wires and other conductors through which the firing is to be done must be known beforehand, so that a battery of sufficient capacity can be selected.

TESTING STRENGTH OF BATTERIES.

Batteries often fail to fire blasts because they can not send such a current as will fire through the great length of leading wires, connectors, and detonator legs used for the blast. A simple way to test the strength of the batteries is to pass the current through a small electric lamp of known capacity (see Pl. XII, *B*) and note the brightness of the light given by the lamp. Another way is to pass the current from the battery through a testing circuit that has in it one electric detonator and whose resistance is equal to that of the circuit of a blast. If the battery fires this detonator (which should be put in a safe place), it is strong enough and is in good condition. (See Pl. XII, *A*.)

USE OF ELECTRIC LIGHT OR POWER CIRCUIT FOR FIRING BLASTS.

In excavation work requiring heavy blasts and the firing of many holes at the same time, it has been found in several instances that the current supplied by ordinary firing machines is inadequate and that misfires often occur. The use of an electric lighting or power circuit in firing electric detonators, when connected in parallel, has in many instances proved advantageous in reducing the number of misfires. This method has been used rather extensively in blasting in the Canal Zone.

The installation necessary for firing shots in parallel by a power circuit is expensive and is warranted only when the work is of sufficient magnitude. For temporary blasting and when only a few shots are to be fired at the same time, the connecting of electric detonators in series and firing by firing machines is more economical.

ELECTRIC FIRING MACHINES SHOULD NOT BE OVERLOADED.

The capacity of firing machines should be determined at frequent intervals in order that they may not be overloaded when used. The failure of shots in blasting operations is often due to overloading, but another reason for failures is the improper operation of the firing machine.

PROPER MANIPULATION OF FIRING MACHINE.

To insure the detonation of all charges the ratchet bar of the machine should be pushed down quickly and forcibly in order to obtain the maximum speed of the revolving armature. Tests made by the

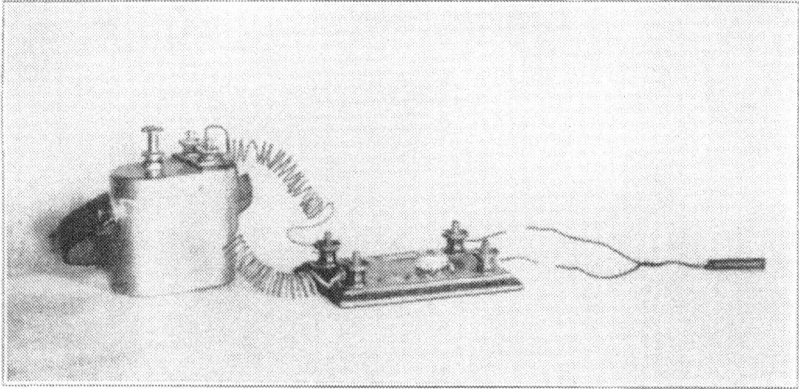

A. TESTING FIRING MACHINES AND BATTERIES.

B. SMALL ELECTRIC LAMP USED FOR TESTING BATTERIES.

bureau showed that when four electric detonators were connected in series and fired with a four-hole firing machine the time interval between the firing of the first and last electric detonator was sometimes 0.005 second. This interval is greater than the time required for a column of low-grade dynamite 30 feet in length to detonate, hence under these conditions the only advantage resulting from using more than one electric detonator in a long charge would be to insure the complete detonation of the charge. When a 10-hole machine was used, all other conditions being the same as in previous tests, the time interval was only 0.0001 second, so that detonation was practically instantaneous. In blasting operations, when more than 10 shots are to be fired at one time, the firing can be best accomplished by wiring all electric detonators in parallel and using a light or power circuit for firing.

FIRING FROM ELECTRIC LIGHTING OR POWER CIRCUIT.

When the electric current for firing is obtained from an electric lighting or power circuit the connections should be made in parallel; that is, one leg of every detonator should be connected to one of the leading wires and all the other legs to the other leading wire. Care must be taken that, by the use of insulating tape, there shall be no short-circuiting in the connections. It is to be borne in mind that in this and other methods of electric firing an accidental premature blast may possibly be caused by leakage from the electric main to the earth and then through the leading wires or connections, which may have become bared by rough handling or may not have been properly covered by the insulating tape. Premature explosions have been known to be caused by leakage due to defective insulation.

DYNAMO-ELECTRIC MACHINES.

Firing machines, sometimes called blasting machines, generate an electric current which is used in firing blasts. Several such machines have been invented, but the two best-known classes are the dynamo-electric machines and the magneto machines. The dynamo-electric machines are made like ordinary dynamos used for generating electric currents, differing only in that they are worked by hand. They contain a coil-wound armature, which is rotated between the poles of an electromagnet. This armature can be made to revolve by means of a crank or a vertical ratchet geared direct to the spindle of the armature. The machines with ratchet bars are so made as to store the current during a stress, until, just as the stroke is ended, the entire current that has been gathered is discharged through the leading wires.

RATING OF DYNAMO-ELECTRIC MACHINES.

These machines are built in different sizes and are rated according to the number of electric detonators they can fire. The machines usually built for use in coal mines are rated as " four-hole " machines, and such a machine can be conveniently carried about by the miner or shot firer.

MAGNETOS.

The magneto machines consist mainly of an armature revolving between the poles of a set of permanent magnets. They look much like the dynamo-electric machines that are worked by cranks, and they are used in much the same way. These magneto machines are used to a considerable extent in foreign countries, but the " pushdown " dynamo-electric machine is the machine most commonly used in the United States.

LEADING WIRES FOR FIRING MACHINES.

The leading wires that carry the current from the blasting machine to the blast hole are insulated copper wires, copper being used because it is one of the best conductors of electricity known and because it possesses the further advantage of becoming little corroded in damp mines. These leading wires, or firing lines, are insulated with a braided covering, which is better when made of waterproof material. In some instances the two wires are twisted together and wrapped with an additional coating of braid, the two wires being thus made into one cable, which has the advantage of being more easily handled than two separate leading wires, and has the added advantage of the protection given by the additional braiding.

INSPECTION AND REPAIR OF LEADING WIRES AFTER FIRING.

After the blast has been fired the ends of the leading wires should be immediately disconnected from the posts of the firing machine and the lines should be examined throughout their whole length in order to see that the insulation has not been broken by coal or rock thrown against it, and has not been stripped by the force of the blast. When such defects or injuries are found they should at once be repaired with insulating tape and then the leading wires should be placed where they are not liable to further injury before being needed again.

TESTING THE FIRING LINE.

To test the line after it has been connected and before firing with it, in order to show that the circuit is complete and that there is no leakage in the wires, a special galvanometer may be used, together with a battery such as many of the manufacturers of explosives now sell. This galvanometer, like others, bears on its face a needle, which is turned or deflected when an electric current is present in the system. By noting whether this needle is or is not deflected one can tell whether the circuit is open or closed, and the extent of the deflection shows just what resistance there is in the circuit. To use the galvanometer, the wires leading from it are connected to the two binding posts of the firing machine to which the wires leading to the charge have already been connected, and the deflection is then noted. The current generated by the weak battery cell attached to the galvanometer should not be strong enough to fire the electric detonator used in the bore holes, but should be strong enough to deflect the galvanometer needle.

SAFETY PRECAUTIONS TO BE TAKEN.

The testing galvanometer, with its attached battery, should never be applied directly to the face to be blasted, even when it is being used to find out which of the electric igniters or electric detonators are defective, after the test has shown no current. The tests for the separate detonators or igniters should always be made through leading wires sufficiently long to allow the person making the tests to stand where he will be perfectly safe in case the blast should be fired, and on no account should this testing of the igniters or detonators be made while any person is so near that he may be in danger from the blast.

THE USE OF EXPLOSIVES IN EXCAVATION WORK.

Practically every kind, class, and grade of explosive is used in open-work blasting. In such work the efficiency of the explosive is the most important factor, its liability to evolve poisonous gases being of secondary importance, and its liability to ignite gas or dust needing rarely to be considered. The use of explosives in sinking shafts and driving tunnels, which is closely allied to their use in deep workings, is discussed in another section.

IMPROPER SELECTION COSTLY.

Explosives used in blasting are frequently bought solely on descriptions of them, though the descriptions may give little or no information concerning their nature, properties, or characteristic components. This oversight accounts largely for the use of explosives of a kind or in a quantity not suited to a given piece of work. Energy and money have often been wasted in blasting operations because of the use in them of expensive, high grade explosives when cheaper, low-grade explosives would have been more effective and more economical, provided that they had been used in the proper location, had been sufficiently stemmed and tamped, and had been fired by sufficiently strong detonators.

VARIOUS EXPLOSIVES FOR USE IN EXCAVATION WORK.

As a guide in making a proper selection of explosives for use in excavation work the following general recommendations are given:

If the texture of the material to be excavated is very tough and hard, as in tough granite or hard bowlders, 60 per cent "straight" nitroglycerin dynamite is recommended for use. If the material is of moderate toughness and somewhat brittle, 50 per cent "straight" nitroglycerin dynamite is recommended. For hard earth or compact sand a 30 to 20 per cent "straight" nitroglycerin dynamite is recommended. In material such as a soft, crumbly, or seamy rock that requires a stronger explosive than black blasting powder, but a slower explosive than dynamite, a granulated nitroglycerin powder containing 5 per cent of nitroglycerin is recommended.

For very soft material in cuts and fills or for quarry work when dimension stone is sought, black blasting powder is recommended. In grading work the blast hole may be chambered with dyna-

54

mite before being charged with granulated nitroglycerin powder
or with black blasting powder, but before it is charged care should
be taken to make sure that the dynamite charge has not left any
fire in the hole. In "plastering" or "adobe" work on bowlders and
spawls, or in "block holing," a strong dynamite should be used.
"Block holing" is the more effective and economical method for use
with bowlders.

AMMONIA DYNAMITES.

If the "straight" nitroglycerin dynamites recommended above
are found to be too quick or too violent for use and the results obtained are not such as are desired under the given circumstances,
ammonia dynamites, which give more of a heaving and rending
action, are recommended. They are made in several grades and are
rated as of a certain percentage of strength, but this rating is not
always made in a scientific way.

DRIVING RAILROAD CUTS.

In driving a railroad cut in comparatively soft material black
blasting powder or a low-grade explosive is usually loaded in the
drill holes after they have been chambered with dynamite, for
on firing the charge the ground is thoroughly loosened, and can then
easily be handled by steam shovels. In this work the holes are always
drilled below the grade stakes in order that the removal of the dirt
by the steam shovels may be expedited. When rock is encountered
and the cut is unusually deep the material is sometimes blasted and
excavated in two or more benches. A higher grade explosive is
used than is used in earth excavations, the row of drill holes driven
is placed closer to the face, and these drill holes are put nearer together. The bottom of the drill holes is not chambered to the
same extent as when black blasting powder is used, because it is
desirable in such work to have the explosive charge extend well up
in the drill hole, for the charge then breaks the rock more completely
and reduces the number of large fragments that might be produced.

DETAILS OF A LARGE BLAST AT A RAILROAD CUT.

In excavation work of magnitude it is sometimes advantageous to
drive a small tunnel into the bank and to load this cavity with explosives. Following is a description of a large blast of this character made at a railroad cut in 1911. (See Pl. XIII, frontispiece.)

The purpose of the blast was to remove a large mass of hard rock
consisting of a pegmatitic and biotitic granite that stood in the form
of a point or nose about 55 feet high and 100 feet across. On ac-

count of the contour of the hill it was impracticable to erect 6-inch
well drills on top and follow the usual procedure in quarry blasting,
as was desired, and therefore the tunnel method was adopted and
carried out as follows:

PRELIMINARY EXCAVATION INTO THE ROCK MASS.

A small tunnel, or drift, varying in width from 4 to 5 feet and in
height from 6 to 7 feet, was driven into the side of the hill to a distance of 80 feet, and at the upper end of this drift one crosscut was
run 65 feet to the right and one crosscut 26 feet to the left (fig. 3).

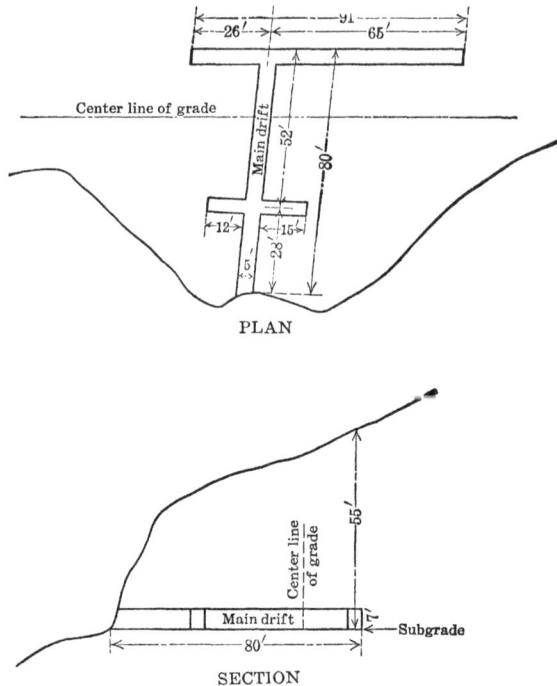

FIGURE 3.—Method of tunneling under large mass of rock before blasting.

Two additional crosscuts were then driven near the mouth of the
main drift, one 15 feet and the other 12 feet, as indicated in the figure.
The cross section of the excavation averaged 35 square feet in the main
drift and 24 square feet in the crosscuts. In driving these drifts compressed-air drills of 1¼-inch diameter were used to bore the holes, and
No. 2, 40 per cent strength, low-freezing dynamite, fired by electricity,
was used to bring down the rock. The tunnel as thus made was
fairly dry, though there was a little water leaking through the roof
in a few places. As the rock in the main drift and in the left-hand
crosscuts was much harder than in the 65-foot right-hand crosscut,
it was decided to use a granulated powder containing 5 per cent of

nitroglycerin in the 65-foot right-hand crosscut and to use a 60 per cent nitroglycerin dynamite in all other places.

The dynamite was taken out of its boxes and the cartridges loaded into place without removal of their paper wrappers. The granulated nitroglycerin powder was loaded into place in the original 12½-pound paper bags in which it was received. Ten No. 8 electric detonators, each containing 2-gram charges of 90 per cent mercury fulminate

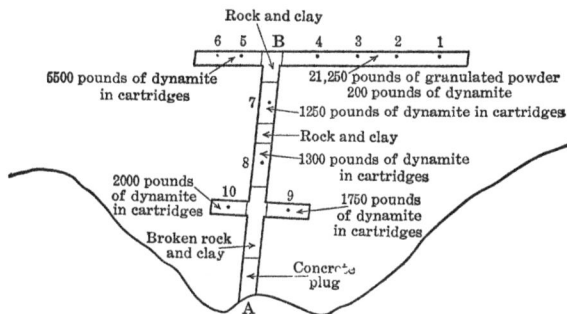

PLAN SHOWING METHOD OF LOADING

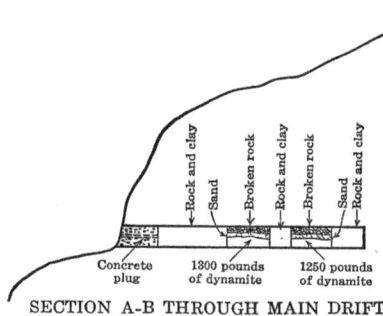

SECTION A-B THROUGH MAIN DRIFT

FIGURE 4.—Method of placing charge in tunnel for blasting large mass of rock.

and 10 per cent potassium chlorate, but having different lengths of No. 20 wires according to location, were used for firing the blast. All connections were made on the outside of the tunnel. The detonators, indicated by the numerals 1, 2, 3, 4, in figure 4, were each placed in 15 pounds of 60 per cent strength low-freezing dynamite, each of the four 15-pound lots being used as priming charges with which to completely detonate the granulated powder.

On the following page are given the quantity and the cost of the explosives and detonators used.

Quantity and cost of explosives, detonators, and electric detonators used in large blast.

12,000 pounds of 60 per cent strength low-freezing dynamite, at 12½ cents per pound_____	$1,500.00
20,250 pounds of 5 per cent granulated nitroglycerin powder, at 7 cents per pound_____	1,487.50
Total_____	2,987.50
1 140-foot No. 8 electric detonator_____	.69
1 120-foot No. 8 electric detonator_____	.59
3 100-foot No. 8 electric detonators_____	1.46
2 80-foot No. 8 electric detonators_____	.78
1 60-foot No. 8 electric detonator_____	.31
3 45-foot No. 8 detonators_____	.64
	4.47
Discount of 25 per cent and 5 per cent_____	1.29
Net cost_____	3.18

MISCELLANEOUS DETAILS.

The legs of the detonators were long enough to extend beyond the bore holes and all wires passed out through a 2-inch iron pipe, 20 feet long, that was embedded in the concrete plug used in stopping the tunnel. An effort was made to reduce the air spacing in all drifts to a minimum by filling with sand and broken rock all spaces not occupied by the explosives. Two weeks was required in loading and tamping the explosives and in building the solid concrete plug at the entrance of the main drift, the leading wires from the detonators being brought through the iron pipe inclosed in the plug. Before the shot was fired the resistance of each of the detonators was measured, and after they had been connected in series the galvanometer showed a total resistance of 32 ohms. Four hundred feet of No. 14 leading wire was then connected to the leading wires at the portal and the resistance of the entire circuit was found to be 36 ohms. A No. 4, 50-hole, push-down firing machine was then tested by firing one electric detonator which was connected in series with a testing coil with a resistance equal to that in the circuit of the blast, and, having been found efficient, this machine was used in firing the blast.

At 2.02 p. m., after all persons present had retired to a place of safety, the shot was fired. The shot firer, protected by a concrete arch, was 400 feet away. When the shot was fired earth or air waves were hardly noticeable. Hence the shot was probably well proportioned. The smoke from the blast was rather dense and had a strong odor of hydrogen sulphide.

ANALYSES OF GASEOUS PRODUCTS OF COMBUSTION.

Immediately after the blast an air sample was taken by displacement at the point where the firing machine was placed, and two minutes after the blast another sample was taken from a crevice from which smoke was coming near the place where the 15-foot right crosscut was driven. Two minutes later another air sample was taken in the same way on a line with the 26-foot left crosscut. The analyses of the three samples gave the following results:

Results of analysis of gaseous products of combustion.

Constituent.	Sample 634.	Sample 635.	Sample 636.
Carbon dioxide..........................	0.30	0.44	2.47
Oxygen.... .. ..:..............	20.51	20.13	18.85
Carbon monoxide......................	.05	.16	.91
Nitrogen...	79.12	79.16	77.06
Methane...			
Hydrogen sulphide.............. ..	(a)	....	(a)
Hydrogen..............................	.02	.11	.21
Ethane................................			
	100.00	100.00	100.00

a Less than 0.02 per cent. Odor perceptible.

RESULTS OF BLAST.

The blast was entirely successful, a complete and rapid detonation having been obtained. Few, if any, large pieces were thrown to a distance of over 400 feet. A few trees were uprooted. An old schoolhouse about 250 feet away was demolished by a falling rock and several thousand yards of bowlders and rocks were thrown down below the grade line. The engineer in charge estimated that 19,500 yards of rock had been moved. The blast had taken out more rock in the back than was expected, but there was solid rock left at the end of the 65-foot crosscut in which the granulated 5 per cent nitroglycerin powder was used.

SUBMARINE BLASTING.

The selection of an explosive for submarine work is of especial importance because of the wetting and low temperature to which the explosive is subjected and because the great expense involved in the preparation of each blast makes a failure serious. It has been generally recognized that any explosive does less useful work when used for blasting under water than it does when used for excavating on land. This lessened effect is due to the fact that the temperature of the products of the explosion is reduced by the chilling effect of the surrounding water and that the pressure of the water on the material to be blasted more than counteracts the beneficial effects of the

water as a stemming.   The procedure usually followed in submarine
excavating is to drill the holes in the underlying rock by means of
machinery placed on a drill boat that is moved about as desired by
adjustable legs on the boat and mooring lines attached to the shore.
The steel drills are guided by long iron pipes, which are usually 4
inches in diameter, and are so lowered from the side of the boat as
to rest on the underlying rock.   The sediment and drillings that
accumulate in the 4-inch pipe during drilling are at intervals removed from it by means of a goose-neck iron pipe 1 inch in diameter.
After the holes have been drilled to the required depth they are loaded
by lowering charges of dynamite into them through the 4-inch pipe.
After the holes have been charged and the primer containing the
electric detonator has been placed on top of the charge, the iron
pipes are carefully removed and the wires from the detonators are
connected in series.   The drill boat is then moved away to a safe
position and the leading wires are connected to a push-down firing
machine.   Waterproof electric detonators should be used in work of
this kind, so that the detonating composition in them may be protected when submerged under water.   In this kind of work, and in
all other blasting operations in which electric detonators are to be
fired in series, it is important to provide a firing machine of adequate capacity and to operate it in such a manner that it will yield
its maximum energy.

ADVANTAGES OF "STRAIGHT" NITROGLYCERIN DYNAMITE.

The results of tests made by the bureau indicate that "straight"
nitroglycerin dynamites are more suitable for submarine blasting
than any other type of explosives, provided the paper wrappers on
the cartridges are not distorted or broken in loading, and provided
also that the charges are not submerged under water for a period
longer than one hour.   The "straight" nitroglycerin dynamites are
also more likely to detonate than are other types of nitroglycerin explosives when the temperature of the water is such as to cause the
nitroglycerin to freeze.   When the temperature of the water is above
the freezing point of nitroglycerin and it is necessary for the charge
of explosive to remain under water for several hours, the gelatin
dynamites have been found to be more suitable for use than
"straight" nitroglycerin dynamite, and under these conditions gelatin dynamite will develop greater energy and do more work than
will "straight" nitroglycerin dynamite.

BLASTING IN CITIES.

The common methods of using explosives in quarrying and open
work are those in general used in making excavations for cellars,
sewers, and trenches in cities and towns, and to some extent for

grading. However, in the crowded communities where the blasting is on or adjacent to thoroughfares and near inhabited buildings, especial care must be exercised in so placing a shot, limiting the quantity of explosive used, and confining it that when fired the shock from the shot shall be almost imperceptible, shall do no damage to adjacent structures, and inflict no discomfort on their inhabitants or on passers-by. Moreover, the blast must be so covered that none of the fragments from it and no part of the system can be thrown so as to harm persons or property.

No specific rules as to the quantity and kind of explosives to be used can be given to meet all of the great variety of conditions encountered in practice. For each new project in untried rock or hardpan a test hole of 2 feet deep containing one-half a stick of 40 per cent " straight " dynamite, stemmed to the collar, well tamped, and thoroughly covered, may be tried, and from the results the plan of procedure as to the depths of holes, the number of holes in a shot, and the size of charge for each hole may be evolved.

The material used as a cover for the blast naturally differs with the circumstances of the locality. As the cover will be subject to hard usage and liable to destruction, it must be cheap. Usually such waste material as old lumber, timber, railroad ties, and brushwood may be used, the lighter materials presenting the larger surface being placed directly over the blast and weighted with the heavy timber or ties placed upon them. It is to be borne in mind that advantage may be taken of the resistance that the atmosphere offers to the movement of bodies through it by choosing for the units of the covering material those with as large a surface as possible, by weaving these units together so that the cover will move as a whole, and by making the surface of the cover as large as practicable, so that it will press against a large surface of the atmosphere. The covering must be of such material, so woven together, and so placed that none of the pieces thrown off from the rock by the blast shall be thrown through or around it.

BLASTING A BUILDING SITE IN PITTSBURGH.

Ensign O. M. Hustvedt, United States Navy, in a recent report to the Bureau of Mines, has described blasting operations carried out May 22, 1914, at Liberty Avenue near Fortieth Street, Pittsburgh, as follows:

Blasting was done during excavation work preparatory to laying foundations for an addition to a small factory building. Excavation was to be carried to a depth of 8 feet. A layer of moderately hard, stratified blue shale was encountered at a depth of 2 to 3 feet, running under the building and supportings its foundations. The character of the layer necessitated great care and the use of small charges, especially at the side of the excavation nearest the foundation of the building.

Bore holes were drilled or churned by hand, a short drill and sledge hammer being first used, then a longer drill and hand power only. The holes were run to a depth of nearly 4 feet and the centers spaced about the same distance. The holes nearest the building were about 3 feet from the foundation line.

A dynamite of 40 per cent strength was used, the charge for each hole being two cartridges (1¼ by 8 inches), the primer being uppermost. Electric detonators, supposedly No. 6, with 4-foot legs, were embedded in the primer from the top, and the two holes were connected in series and fired simultaneously. A push-down blasting machine of the usual type was used for firing the charges. The stemming used was ground mortar from dismantled brickwork, and was tamped by hand.

A feature of especial interest was the means used to prevent débris from flying when the blast was fired. A number of old wire bed springs, with closely woven tops and bottoms, separated by helical distance springs, were piled in several layers over an area of about 6 by 12 feet, with the two bore holes approximately at the center. Over the bed springs were placed four or five heavy railroad ties to weight them down. When the blast was fired the railroad ties were lifted to a height of 5 or 6 feet, but the springs moved very little, and no fragments escaped from under them, so far as could be observed. The protection afforded by the springs seemed very effective, as was to be expected, as the wire mesh was fine enough to retain very small fragments and yet to exert no confining effect upon the gases.

The springs are practically indestructible under the conditions of use and their lightness and their ease in handling also recommends them.

The effect of the blast upon the rock was seemingly satisfactory, two holes charged as above breaking up about 5 or 6 cubic yards into fragments of nearly uniform size in fairly good shape for handling.

USE OF EXPLOSIVES IN QUARRYING.

CONDITIONS AFFECTING CHOICE OF EXPLOSIVES.

The methods employed in quarrying vary greatly in different quarries because of the wide differences in chemical and mineralogical composition and physical constitution of the rocks, and still more because of the use to which the product is put. In the past, low explosives and, to some extent, high explosives, such as dynamite, have been used in the quarrying of dimension stone to be used in buildings or for monuments, but it is believed that instances of deterioration of such stones observed in use have been due to minute cracks and fissues developed in them by the shock from the explosive used, and the tendency to-day is to employ channeling machines or plugs and feather wedges, or, in some instances, compressed air for blocking out this material.

INCREASED USE OF BROKEN ROCK.

However, a continually increasing quantity of broken rock is now being used as flux in metallurgical operations, as ballast for railroads, as road metal, in the manufacture of cement and of concrete, for rubble at breakwaters, and for other purposes for which the shattering effect of such explosives as may be used in blasting the rock does not harm the rock for use, and hence blasting operations for the production of stone for such uses are now of wide extent.

STRIPPING.

In almost all instances the rock mass to be worked is overlaid with earth or weathered rock, and this overburden must be stripped off to give access to the workable deposit. Such stripping is often most easily accomplished by blasting, and this method is generally followed.

BENCH METHOD OF QUARRYING.

In quarrying in the past the bench method, with drill holes of small diameter, which in many rock masses required chambering at the bottom before the main charge was loaded, was used.

FACE METHOD.

At present, especially where ballast or rock for concrete work and quarried ore or flux are sought, the method of blasting by large drill holes is being rapidly introduced. In some quarries a line of 6-inch holes each 100 feet in depth is drilled and several thousand pounds of explosive is used in a single blast. The charge usually extends 30 feet up from the bottom of the hole. It has been found that when one electric detonator is placed in the top of one of these long charges it does not insure a complete detonation of the entire charge of explosive in the drill hole; therefore, two or more connected electric detonators distributed through the charge are generally used. When the most violent effect is desired in blasting, the best method of placing electric detonators in a charge 30 feet in length, whether they are connected in series or in parallel, is to place one electric detonator 5 feet above the bottom of the charge, one 5 feet below the top of the charge, and one in the center of the charge. If it be assumed that the entire charge detonates at a uniform rate and if the three electric detonators are fired simultaneously, the duration of the explosive reaction will be one-sixth of the time that would elapse if only one electric detonator were used in the top of the charge. Electric detonators are to be preferred, because ordinary detonators, widely separated, might not explode at the same time. Souder[a] states that the cordeau detonant is especially efficient and safe for use in the damp deep holes of the open-cut iron mines of Cornwall, Pa.

QUARRYING IN UTAH.

The use of large and deep drill holes is practiced in quarrying a rather soft rock (monzonite porphyry) in Utah. Deep holes are sunk with churn drills and loaded with large charges of dynamite. The holes are $6\frac{1}{2}$ inches in diameter and are drilled to a depth of 150 feet. The bottom of each hole is drilled to a point a few feet below the level of the floor of the quarry so as to insure that, after the blast has been fired, there will be loose ground below the level on which the steam shovels work, the work being thus made easier. The holes are drilled about 30 feet apart and 30 feet distant from the face at the bottom. From 6 to 9 holes are shot at one time with 1,200 to 4,700 pounds of dynamite per hole. Both 40 and 60 per cent nitroglycerin dynamites are used, the former grade being more commonly employed.

[a] Souder, Harrison, A new safety detonating fuse: Bull. Am. Inst. Min. Eng., October. 1914, pp. 2547–2556.

LARGE BLAST IN QUARRY AT TENINO, WASH.

In certain quarrying work it has been desired to produce several tons of stone in one blast, and to do this a small tunnel has been driven into the face of the quarry and loaded with high explosives. Following is a description[a] of a large blast of this kind that was made on February 17, 1912, in a quarry at Tenino, Wash. The object of the blast was to produce several thousand tons of sandstone for the Government jetty at Grays Harbor, Wash. The rock was a greenish sandstone of massive structure and with a medium fine grain and stood in the form of a point about 500 feet wide and 50 to 70 feet high at the face. From the top of the quarry face the rock sloped gently upward until at a distance of 200 feet it had an average height of 75 feet above the quarry floor. The mass was cut off on each side by gulches.

In order to blast the rock two main tunnels, each with a number of crosscuts, were driven to serve as chambers for the explosives. The tunnels were about 3 feet by 4 feet in section and were driven with compressed-air drills. The location and length of the tunnels and the arrangement and quantity of explosives and of stemming used are shown in figure 5.

One thousand seven hundred and twenty-four kegs, or 43,100 pounds, of blasting powder and 1,200 pounds of " 60 per cent " dynamite were used in the work. In arranging the charges the black powder was left in its original kegs, the bung of every fifth keg being left open. In the center of each black-powder charge a priming charge of dynamite was placed, about 1 pound of dynamite being thus used to every 50 pounds of black blasting powder. The charging of the powder required 6 days.

Both muck and timber were used for confining the charges. Fifteen bags of sacked muck was laid against the kegs of powder, and muck was then shoveled in until the crosscut was filled to within about 6 feet of the main drift. The remaining space in the crosscut was filled with 6 by 8 inch ties, laid crosswise. As the crosscuts were opposite each other in the main drift, the ties in them could be wedged in by cross ties in the main drift. Both of the main drifts were filled from the last crosscut to the mouth with muck, a cement-and-timber plug being used to close the entrance. Fifteen bags of Portland cement were used for this purpose in each of the two drifts.

A double system of wiring was used, as shown in figure 6. There were fourteen distinct charges, and two No. 6 detonators, each containing 15.32 grains of fulminate of mercury, were placed in each

---

[a] From Mining and Engineering World, Great blast in quarry at Tenino, Wash.: vol. 36, Mar. 30, 1912, p. 719.

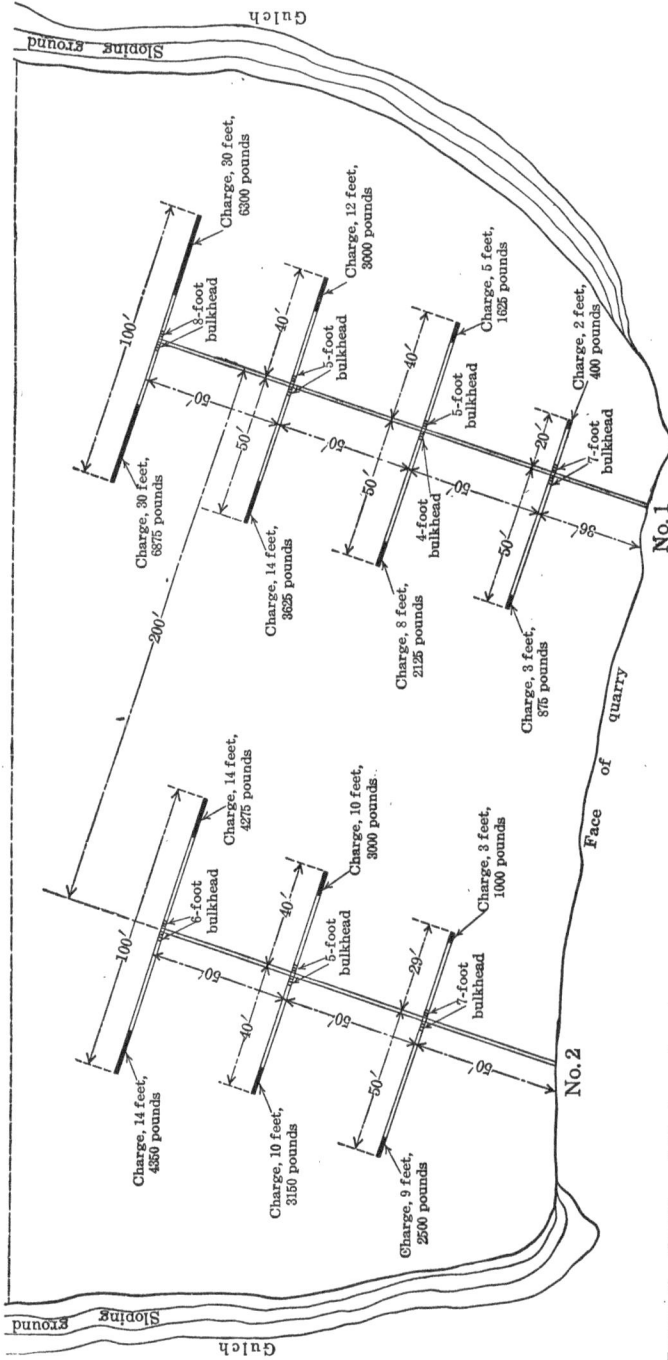

FIGURE 5.—Position of black-powder charges and of timber stemming in two tunnels driven in a rock quarry at Tenino, Wash. Both tunnels and the remaining space in the crosscuts were completely filled with muck stemming. The primer charges of dynamite were placed in the middle of the black-powder charges.

charge. Each detonator carried 30 feet of wire for connections. The resistance of each detonator was about 3 ohms. The detonator legs were each connected to a waterproof, 20-gage copper wire which was carried to the mouth of the tunnels, where connections were made with 16-gage wires running to a push-down battery of 45-detonator capacity, which was placed 1,200 feet away. The wires of both circuits in the tunnels were laid together in a wooden trough made of 2¼ by 4 inch cedar scantling cut with a groove 1¼

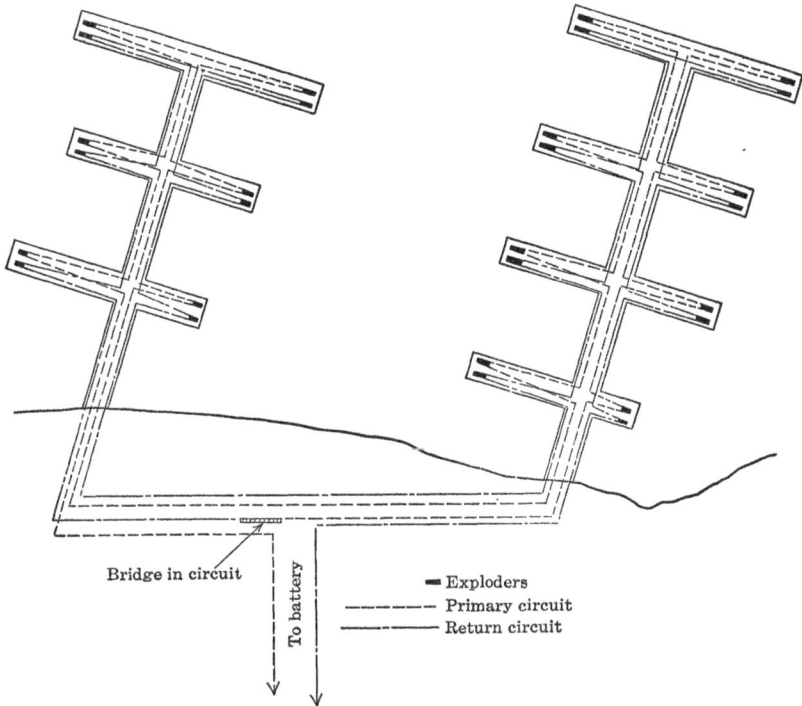

Bridge in circuit

To battery

━ Exploders
--------- Primary circuit
————·— Return circuit

FIGURE 6.—Diagram of wiring at Tenino, Wash.

inches square. A batten was then placed over the groove and nailed to the scantling.

Before the shot was fired all connections were tested with a galvanometer. Before being used the push-down battery was also tested with a special rheostat. The result of the shot was considered entirely satisfactory by the officials of the company. It was estimated that about 220,000 cubic yards of rock was disturbed by the blast. As 43,100 pounds of black blasting powder and 1,200 pounds of "60 per cent" dynamite were used, 1 pound of explosive for about each 5 cubic yards of rock disturbed was required. Several large blocks of stone were produced, and they were subsequently broken by the block-holing method.

As examples of large-scale quarrying operations the Porto Bello and Ancon quarries, operated by the Government in the Canal Zone, may be cited.[a]

QUARRYING AT THE PORTO BELLO QUARRY.

The Porto Bello quarry is about 20 miles east of Colon, in the Republic of Panama, just north of and bordering the entrance to the harbor of Porto Bello. The rock quarried is basalt and andesite. The quarry was opened in 1908 and crushing commenced on March 2, 1909, for the production of crushed rock for the Gatun locks and the Gatun spillway, and continued in operation to April 30, 1912, when certain changes in the equipment were made in order that it could be used for the production of large rock for the protection of the Colon breakwater. This production of large rock began on August 18, 1911. Excavation was effected by steam shovels, but during the development period considerable hydraulic stripping and some hand excavation was done. The rock was loaded into cars, transported to the crushing plant, and dumped directly into the crusher. The crushed-rock quarry was first operated on a climbing cut from which benches were later developed. On June 30, 1910, the face of the quarry was 2,500 feet long with a maximum height of 140 feet. A year later the face was 2,600 feet long with a maximum height of 170 feet. The breakwater quarry was first developed in two benches on a level lower than the crushed-rock quarry. Each bench had a maximum face of 60 feet. The lower bench was 1,100 feet long on June 30, 1912. Later the larger part of the large rock was obtained from the upper level.

For the crushed-rock quarry the drilling was done with $3\frac{5}{8}$-inch tripod drills. Vertical, toe, and breast holes were used, spaced 6 to 14 feet apart, the vertical holes having a maximum depth of 24 feet. The explosive generally used was a " 60 per cent " dynamite, with 15 to 50 pounds per hole, the average charge being larger in the toe holes than in the " down " holes.

In the upper level of the breakwater quarry, where it was desired to produce only large pieces of rock weighing 2 to 20 tons each, the drilling was done with tripod drills. Toe and breast holes were used. The holes were spaced 9 to 15 feet apart and had a maximum depth of 24 feet. Both " 45 per cent " dynamite and " 45 per cent " Trojan powder were used, the usual charge being 40 to 60 pounds per hole.

Previous to September 4, 1910, it was difficult and expensive to reduce the rock to a size that would enter the largest crusher, a No. 9,

---

[a] Contributed by Spencer P. Howell.

because of the great amount of adobe blasting necessary, but on that date a No. 21 gyratory crusher was put in operation, and since then adobe blasting has been little used, the capacity of the plant has been largely increased, and the cost of producing crushed stone has been correspondingly reduced.

There was produced and shipped from the crushed-rock quarry a total of 1,921,575 cubic yards. Available statistics [a] showing the cost of blasting 1,709,365 cubic yards of this material gave a cost figure of $0.2118 per cubic yard. The electric detonators used in all primary blasting were connected in parallel and fired by means of a direct current of 110 volts.

QUARRYING AT THE ANCON QUARRY.

The Ancon quarry is situated on the west side of Ancon Hill and overlooks the site of the new wharves at Balboa. The crushing of the rhyolite rock that it supplies was begun on February 8, 1910, and up to April 1, 1913, a total of 2,412,951 cubic yards had been produced. The quarry supplied crushed rock for the Pedro Miguel and Miraflores locks, and for the fortifications, roads, track ballast, and commercial sales. Excavation for tracks and crusher foundations and stripping was done with steam shovels, but during the development period some handwork was also done. The rock was loaded by means of steam shovels into cars, transported over the switch-back tracks erected there, and dumped directly into the crusher.

The quarry was developed in three benches (Pl. IX, *B*) whose faces were approximately 65, 70, and 265 feet high. On April 1, 1913, each bench averaged 200 feet in width, and the length of the working face of the lowest bench was about 1,400 feet. The crusher was erected on the lowest bench and was at an elevation of 175 feet above mean sea level. Both tripod and well (or mechanical churn) drills were used, the former for toe and breast holes and the latter for deep "down" holes. The toe and breast holes were drilled to a maximum length of 30 feet and spaced 8 feet to 15 feet apart. The "down" holes were usually drilled deep, some 75 feet below the surface, and were spaced about 50 feet apart. "Forty-five per cent" dynamite was generally used, but both black blasting powder and "60 per cent" dynamite were also used. In the toe and breast holes the charge varied from 20 to 150 pounds, depending on the length of the hole, the quantity of the overlying material, and the visible faults. The "down" holes were much more heavily charged.

The cost of blasting for the period January 1, 1910, to June 30, 1912, was $0.0488 per cubic yard for a total of 1,870,277 cubic yards

---

[a] Annual reports of the Isthmian Canal Commission for 1910, 1911, and 1912.

A PRIMER ON EXPLOSIVES.

crushed. The electric detonators used in the primary blasting were connected in parallel and fired by means of a 110-volt alternating current taken from a power circuit.

BLASTS AT TOMKINS COVE, N. Y., QUARRY.

An excellent description of two blasts fired on March 25, 1914, for the production of ballast and concrete stone at a bluestone quarry at Tomkins Cove, N. Y., has been given in a recent report to the bureau by A. J. Hazlewood, from which it appears that both the tunnel and drill-hole methods were used simultaneously, the tunnels being driven at the foot of the wall to a distance of 30 to 70 feet, with branches from each side at the end, thereby making a T-shaped opening 3 by 3 feet in cross section. The drill holes were driven in line vertically downward from the top of the cliff and fired simultaneously with the tunnels. The drill holes were used to supplement the tunnels so as to avoid overhanging of the upper part of the rock mass. The explosives used were 37,275 pounds of 60 per cent strength gelatin dynamite and 41,900 pounds of granulated nitroglycerin powder. The disposition of the charges and the number and dimensions of the drill holes used are shown in the following tables:

Quantity of explosives used in drill holes and tunnels of each blast, Quarry No. 1, Tompkins Cove, N. Y.

Location of blast.	No.	Depth of hole.	Quantity of 60 per cent strength gelatin dynamite.	Quantity of granulated nitroglycerin powder.
		Feet.	*Pounds.*	*Pounds.*
Tunnel	15	....	10,150	14,200
Do	7	....	6,250	10,100
Drill hole	1	83	200	....
Do	2	84	250	....
Do	3	85	250	150
Do	4	84	300	150
Do	5	85	300	150
Do	6	83	300	150
Do	7	83	250	....
Do	8	85	250	150
Do	9	84	200	150
Do	10	84	300	150
Do	11	84	300	150
Do	12	84	250	1 0
Do	13	80	250	150
Do	14	75	250	150
Do	15	71	300	150
Do	16	72	100	150
Do	17	73	250	....
Do	18	75	250	150
Do	19	75	250	150
Do	20	84	450	....
Do	21	42	350	....
Do	22	36	450	....
Do	23	35	400	....
Do	24	35	350	....
Do	25	35	300	....
Do	26	35	75	....
Total	....	....	23,575	26,550

NOTES.

The weight of charge for each hole was determined by estimation of work that each was to do.

The estimated quantity of stone blasted was 210,000 cubic yards.

Quantity of explosive used in drill holes and tunnels of each blast, Quarry No. 2, Tompkins Cove, N. Y.

Location of blast.	No.	Depth of hole.	Quantity of 60 per cent strength gelatin dynamite.	Quantity of granulated nitro-glycerin powder.
		Feet.	*Pounds.*	*Pounds.*
Tunnel	8		1,400	1,000
Do	9		1,600	500
Do	10		1,650	3,600
Do	11		1,500	3,000
Do	12		1,650	3,400
Do	13		1,700	3,500
Drill hole	1	60	250	
Do	2	35	350	
Do	3	35	250	
Do	4	36	250	
Do	5	38	250	
Do	6	36	250	
Do	7	35	250	
Do	8	35	250	
Do	9	35	150	
Do	10	35	200	
Do	11	34	250	
Do	12	34	150	50
Do	13	33	200	50
Do	14	33	200	50
Do	15	34	200	50
Do	16	34	150	50
Do	17	33	200	
Do	18	32	200	50
Do	19	32	200	50
Total			13,700	15,350

NOTES.

The weight of charge for each hole was determined by estimation of the work that each was to do.

The estimated quantity of stone blasted was 72,000 cubic yards.

The tunnels were charged with the explosives discontinuously, the spaces between the charges being so filled with fine rock stemming that the loading density of the explosives used became 1. The endeavor in these blasts was to obtain a uniform pressure and crushing effect from the blast; hence those parts of the tunnels that were in soft rock were plugged with cement and concrete before the charge of explosive was laid, and one of the entrances was plugged with cross rails and concrete work.

USE OF EXPLOSIVES IN METAL MINING AND TUNNELING.

As a rule, except in the case of such ore bodies as those of big iron mines of the Mesabi Range, Minnesota, the disseminated copper sulphide or "porphyry copper" mines, or the placer deposits of the precious metals, the metallic minerals sought in mining more usually occur as veins, stringers, or irregular masses in faults, or rifts in rocks, and are generally reached by shafts with drifts, adits, or galleries, or preferably, when advantageously situated, by tunnels from below which facilitate the use of overhead stoping in breaking ore, and of gravity in moving it, and also keep the mine drained of water.

SHAFT SINKING.

The character of a shaft will vary with the material in which it is sunk and the purpose for which it is to be used. Circular test shafts $3\frac{1}{3}$ feet in diameter have been sunk in soft ground to depths of 100 feet and more by the use of the pick and shovel. Ordinary shafts are square and average 4 by 4 feet in cross section. Where it is intended that the shaft shall carry the pumping machinery and the like, in addition to the buckets, they may be 8 by 6 feet in cross section and be, when completed, divided into two compartments.

DANGER IN BLASTING IN SHAFTS.

Where the shafts are sunk in rock, blasting must be done, and it is obvious that blasting must be attended with danger, particularly when burning fuse is used to fire the blast, because the miner, after having lighted the fuse, must escape from the shaft before the charge explodes. This is difficult and the difficulty increases with the depth to which the shaft is sunk. Moreover, he is menaced by any error in judging the proper length of fuse to use or by any defect in the fuse used. Furthermore, to properly ventilate a shaft while it is being sunk is difficult, and it is therefore important that as little as possible of noxious fumes should be produced, and yet burning fuse is a marked source of production of noxious gases. The remedy lies in the use of electric detonators, as they may be fired by means of blasting machines from the surface of the ground after the miners have left the shaft. It is advised, in order to obtain the highest efficiency, that the charges be well tamped in the bore holes and that the strongest detonators procurable be used, as they are more likely to cause the complete explosion of the charge even when it is partly frozen or when it is damp, and therefore reduce the likelihood of noxious gases being developed through the imperfect explosion of the charge.

USE OF DELAY-ACTION DETONATORS.

It not infrequently happens that it is desired to fire a series of shots in the bottom of the shaft with the intent that the explosion of the breaking-down shot shall precede that of the other shots. Morse[a] describes how this may be done by the use of delay-action detonators, as follows:

In blasting in a wet shaft current should be taken from the light circuit, as there are many sources of leakage and the ordinary blasting machine will not supply sufficient current. Two wires are brought from the light circuit to a small box, which should contain fuses of lower amperage than those on the light circuit, a knife switch opening down and held open with a weight or spring, and two binding posts; an indicator lamp is convenient. This box should have a lock for which the blaster only has a key, and should be placed near the collar of the shaft or at a convenient station. Near by is a reel for holding and paying out a No. 10 encased cable. The inside end of the cable is left protruding sufficiently to permit its being connected to the binding posts when the box is opened. The cable reaches to within a few inches of the bottom of the shaft, and to each of its two wires a length of bare telephone wire is attached and placed horizontally across the shaft, being kept off the bottom with blocks of wood. To these wires the exploders are attached in parallel by the insulated wires with which each is supplied. In loading the holes the first delays are, of course, used in the cut holes. For most work five periods of delay are all that are required. When the men, including the blaster, reach the surface the blaster unlocks the box, connects the ends of the cable to the binding posts, and throws the switch.

DRIVING TUNNELS.

The ordinary procedure followed in this country in driving large tunnels through rock is to drive a top heading the full section of the arch cut and to then remove the bench, but in metal mining this system is often not used. Usually 15 to 30 holes are loaded and then groups of 4 or 5 holes fired successively. The first set of shots that are fired in the center of the face is styled as either the square or the V-shaped center cut. The holes generally converge to a point, for the charges are usually fired by " influence," and it is necessary to have these charges in close proximity to one another in order to

---

[a] Morse, C. W., Electric blasting in shafts with delay-action exploders: Min. and Sci. Press, Jan. 31, 1914, p. 216.

insure their simultaneous explosion, especially when fused detonators are used to ignite the charge. The work is then widened out by firing successively the remaining sets of holes that usually surround the center cut, and is finished by firing the "trimming-up" shots.[a]

USE OF DETONATORS FOR FIRING DEPENDENT SHOTS.

It has previously been stated that the practice of firing charges of explosives by electric detonators is most effective in many kinds of work because it enables one to fire all of the shots simultaneously. There are, however, many conditions arising in mining that do not permit this advantageous use of electric detonators being put into practice. For instance, in driving drifts it is often necessary to fire dependent shots, and when this practice is followed, if electric detonators are used, the falling rock from one shot may disconnect the wires of the electric detonators wired for succeeding shots or may cause short circuiting, hence fuse is used in many of the metal mines, different lengths being cut off for use in the successive shots. The projecting ends, before lighting, are coiled and placed within the mouth of the collar of the hole, where they are well protected.

OBJECTION TO FUSED DETONATORS FOR WET BLASTING.

A serious objection to the use of ordinary fuse and detonators in wet blasting is that it is almost impossible to produce a water proof seal at the top of the detonator by crimping it on the fuse. The ordinary fuse crimper (Pl. XIV, A) operates by flattening the sides of the copper shell in such a way as to contract the diameter of the detonator in one direction and extend it in the other, thus making openings alongside of the fuse. Tests made show that when a detonator is crimped on a fuse in this manner and submerged under water for 30 minutes the fulminating charge and the powder train at the end of the fuse in the detonator become damp, and that on lighting the fuse, even if it burns through, the spit produced by it is usually of insufficient intensity to cause an explosion of the fulminating charge, though in some instances an explosion of a very low order occurs. If only a little water enters the detonator, the spit of the burning fuse may be sufficient to cause the fulminating charge to explode with a sharp report and to completely destroy the copper shell, but without doing useful work. In some tests made by the bureau 70 to 80 per cent of the compressed fulminating charge was recovered in the lower part of the copper shell. The spit of the burning fuse had seemingly caused a part of the fulminating composition to explode with sufficient force to destroy the top part of

---

[a] For a fuller description see Brunton, D. W., and Davis, J. A., Safety and efficiency in mine tunneling: Bull. 57, Bureau of Mines, 1914, pp. 131–146.

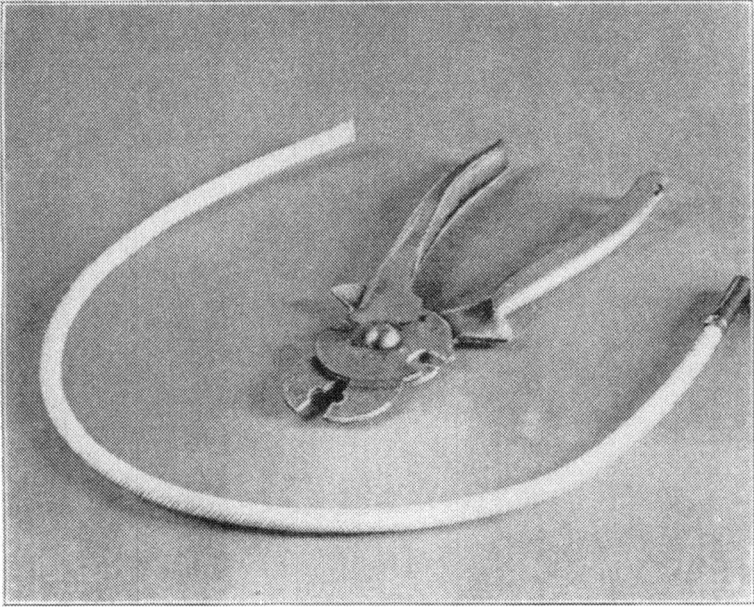

A. ORDINARY CRIMPER.

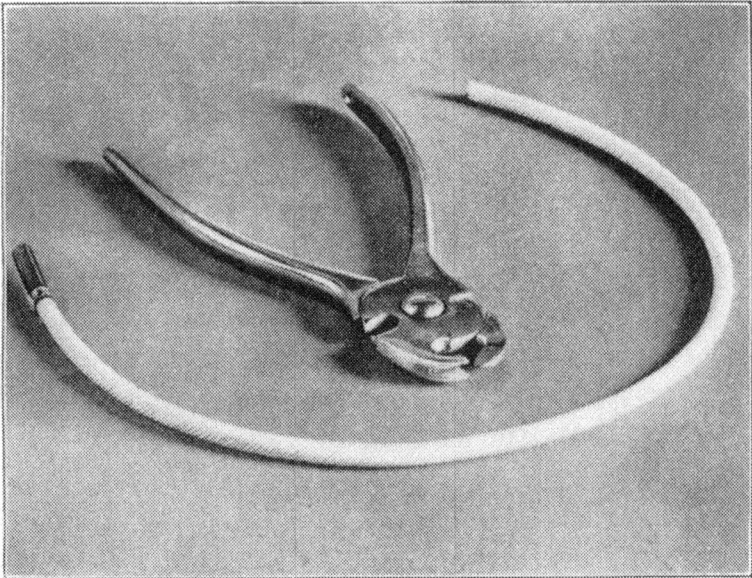

B. NEW CRIMPER.

the copper shell, but no detonation was propagated throughout the remainder of the wet fulminating charge. In these instances only a slight report was audible, and it was obvious that an explosion of this order would not cause a complete detonation of dynamite or other high exposives.

SEALING WITH TALLOW.

In some of the tests a thin coating of tallow was spread on the fuse, one-fourth of an inch from its end and extending one-half an inch up its side, before inserting it into the detonator, and thereby a more nearly perfect seal was made.

USE OF NEW CRIMPER.

A crimper (Pd. XIV, *B*) recently placed on the market by a manufacturer of explosives crimps the detonator on the fuse in a different manner from that of any of the types of crimpers previously used. The salient feature of this crimper is its ability to contract the top of the detonator uniformly and to form a groove, one-eighth of an inch wide, around the copper shell, thus perfecting a seal of the detonator onto the fuse that will permit submersion of the detonator in water for 30 minutes without destroying its efficiency. The zone of the shell is pressed firmly and uniformly into the fuse, but not so far in as to break or separate the powder train in the fuse. Owing to the varying diameters of different types of fuse and the probability of considerable variation in the same type or even the same coil of fuse, the use of a thin film of tallow around that end of the fuse that is inserted in the detonator, as described above, will insure a better seal, irrespective of the crimper used.

USE OF EXPLOSIVES IN LARGE PROJECTS.

Following are descriptions of engineering projects of magnitude, in which different systems were pursued in driving tunnels, in this and foreign countries, and are cited as furnishing detailed information regarding the use of explosives under the varying conditions that arise in practice.

NEW BUFFALO WATERWORKS TUNNEL.

SIZE AND CHARACTER.

The new Buffalo waterworks tunnel [a] consists of two sections—one 12 feet wide, 11 feet high, and 4,270 feet long, and the other 15 feet by 15 feet in section, and 6,575 feet long. In the small tunnel 700

---

[a] See Lavis, F., The new Buffalo waterworks tunnel: Eng. Rec., June 25, 1910, p. 802.

feet of the rock was hard limestone, the remainder being a comparatively soft shale. In the main tunnel 1,800 feet was driven through
limestone and the remainder through a harder rock of a flinty nature.
A considerable quantity of water was encountered and most of the
work was driven under a compressed-air pressure of 15 to 28 pounds.

METHOD OF DRIVING.

The method of attack was by top heading and benches, and the
drills used were $3\frac{1}{4}$-inch compressed-air machines. Four drills,
mounted on two columns, were used in the heading and two tripod
drills on the bench. The method of spacing the drill holes is shown
in figure 7.

EXPLOSIVES USED.

Except that 1,000 pounds of 96 per cent blasting gelatin was tried as
an experiment, the powder used throughout was 60 per cent strength

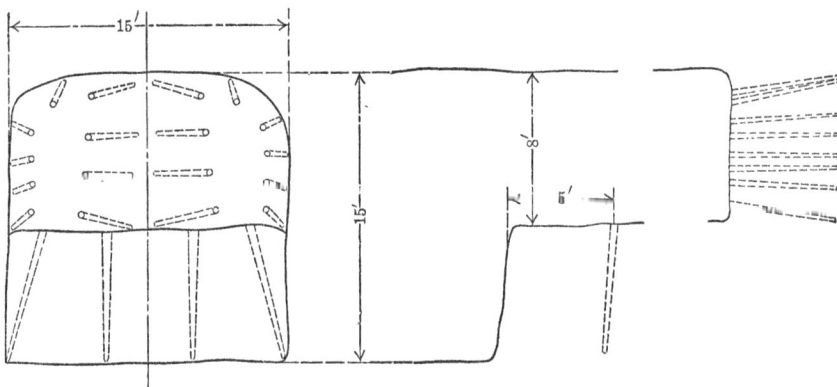

FIGURE 7.—Arrangement of drill holes in face and heading, new Buffalo water-works
tunnel.

gelatin dynamite. The blasting gelatin cost about 8 cents per pound
more than the 60 per cent gelatin, but no advantage resulted from its
use either through its permitting the use of a smaller quantity of explosive or through its breaking the rock better, although careful experiments were made to determine those points. In order to save
time in loading, "sticks" 16 inches in length instead of the usual
8-inch length were used, the bags of stemming used being of the same
length. The cut holes were loaded with 4 sticks (about 6.4 pounds)
and the remainder of the holes with $3\frac{1}{2}$ sticks (5.6 pounds), so that
the quantity used for one round of bench and heading was about 113
pounds. If the average advance be taken as 5 feet, the quantity of
explosive used per cubic yard was somewhat less than 3 pounds. For
comparison it may be mentioned that in the Simplon and Loetschberg tunnels about 7 pounds per cubic yard of heading was used.

Some little trouble was experienced in the early stages of the work when low-grade detonators were being used, as the rock was very cold, and many holes, especially those at the bottom, were filled with water so cold that it chilled the powder. In several instances in the early stages of the work the detonator went off, but failed to explode the charge. No. 8 electric detonators were then tried, and the results were uniformly successful. All detonators before use were tested with a small pocket galvanometer. The economy of using strong high-grade detonators was considered fully proved on this work. Electric delay-action detonators were used with success and resulted in a great saving of time. When detonators alone were used 5 rounds were required to fire all the holes in the bench and heading, and the time of blasting was $1\frac{1}{2}$ to 2 hours. By the use of electric delay-action detonators all holes could be fired in two rounds, and the time of blasting was thus reduced to one-half hour. First delay-action and second delay-action detonators were used, each class of detonator having a different colored insulation on the lead wires in order that they might be readily identified. Each pair of drillers loaded their own holes.

HANDLING OF EXPLOSIVES.

The thaw house in the tunnel was kept warm in cold weather by a steam radiator covered with a screen, the explosive being placed on shelves on the side of the room opposite from the radiator. Each shift laid on the shelves as much powder as it took away.

The powder was taken into the tunnel in a wooden box mounted on the frame of one of the muck cars, with rubber cushions under the box and 8 to 10 inches of excelsior in the bottom of the box. The heading boss brought out all powder and detonators that were not used. It is stated that there were no powder accidents during the entire work.

FIRING SHOTS.

All shots were fired by the current from the electric-light circuit. The blasting line was of No. 14 copper wire, with weather-proof insulation, and was strung on the side of the tunnel opposite the lighting circuit. The switch was 700 to 900 feet back from the heading, and was moved up from time to time as the work progressed. The firing line was tested by the electrician before each shift. The lead wires were never connected until a drill runner had been sent back and stationed at the switch, and the connection to the light line was not made until all hands were away from the face. The latter connection to the nearest convenient socket was made low enough, so that the connecting wire between the socket and switch would not be fouled by a car or man passing on the track; so there was only remote danger in leaving this part of the blasting line connected after the blasting had been finished.

DRILLING AND BLASTING METHODS ON NEW YORK RAPID-TRANSIT TUNNEL (SUBWAY).

The New York rapid-transit tunnel (subway) was driven 28 feet wide by 20 feet high. The method of excavating in this work was that of the single-top heading and bench, usually employed in the United States. The rock drilled through was mainly schist. The approximate location and depth of holes in a blast is shown in the following table:

Data regarding drill holes used in New York rapid-transit tunnel (subway).

BENCH HOLES.

Order of firing.	Number and kind of hole.[a]	Depth.	Size of charge.	Explosive used.
		Feet.	*Pounds.*	
A........	7 grading.....................	3–5	50	"40 per cent" dynamite.
B........	5 bench.....	9½	45	Do.

HEADING HOLES.

B........	6 trimming.....................	3–9	42	"40 per cent" dynamite.
C........	8 center cut....................	9	56	"60 per cent" dynamite.
D........	8 side.........................	8	48	"40 per cent" dynamite.
E........	8 dry..........................	8	36	Do.

a All holes tapered from 3 to 2¼ inches in diameter.

RONDOUT PRESSURE TUNNEL.

The Rondout pressure tunnel[a] is part of the Catskill Aqueduct that is to carry water from the Ashokan reservoir to New York City. The tunnel is 23,608 feet long and was driven with a circular cross section, 17 feet in diameter. Various sedimentary rocks were penetrated, varying from soft shales to limestones, sandstones, and a hard quartz-conglomerate. Ordinarily the rock was easily drilled and the position of the strata was favorable to good progress. No timbering was required and little water was encountered.

The top heading and bench method of driving was used, about half of the tunnel section being taken out with each operation. The bench excavation was kept about 50 feet back of the heading. For

a See Hogan, J. P., Progress on the Rondout pressure tunnel: Eng. Rec., Jan. 1, 1910, p. 26.

drilling the heading four 3¼-inch machine drills, mounted on two vertical columns, were used, and for the bench two tripod drills of the same type were employed. As the men were paid for a fixed section in excavating, great care was taken in placing the holes and there was little excess breakage. The number and position of the drill holes required to blast out a round of heading is shown in figure 8. The average number of holes in a heading was 22, divided into 6 cut holes, 6 side or relief holes, and 10 rim or trimming holes. The cut holes were 10 to 12 feet deep and the remaining holes 8 to 10 feet, depending on the amount of ground to be broken. The bench holes were spaced 4 feet apart and averaged 4 holes. Two rounds of the bench were shot with the cut; the side and trimming holes were then loaded and shot successively, making three shots for a complete advance. The full round required 175 to 200 pounds of explosive, the grade used being a "60 per cent" gelatin dynamite.

This being a pressure tunnel, a minimum of 200 feet of rock cover was calculated to be necessary. The tunnel was driven from shafts, eight being sunk along its line. They ranged in depth from 350 to 710 feet. Three of them are circular in section and will be used permanently in the operation of the

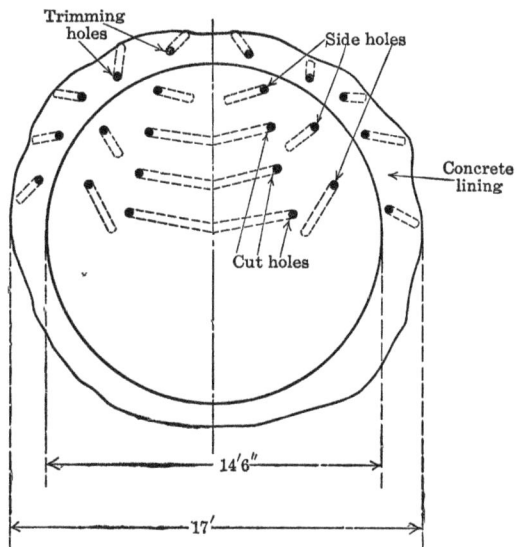

FIGURE 8.—Arrangement of drill holes in heading of Rondout tunnel.

aqueduct. Five are rectangular, 10 by 22 feet in section. During a single month's work in one heading a rate of advance of 488½ feet was made.

HUNTER BROOK TUNNEL.

The Hunter Brook Tunnel[a] is also part of the Catskill Aqueduct and is of interest because it is one of the few long tunnels of America that have been driven with a bottom heading. It is 15 feet wide, 18 feet high, and 6,150 feet long. The rock penetrated was a mica schist and a hard, close-grained gneiss. Numerous faults and slips

[a] See Becker, Arnold, Bottom-heading driving on the Hunter Brook Tunnel: Eng. Rec., vol. 64, Sept. 23, 1911, p. 358.

made the ground treacherous and difficult to work. In driving the heading 3½-inch air drills were used, 2½-inch drills being used for the stope.

In order to facilitate the haulage of muck the bottom heading was driven the full width of the tunnel. The dimensions are given in figure 9. To blast out a heading round required 16 to 28 holes, varying in depth from 5 to 7 feet. The powder used was a "60 per cent" gelatin dynamite, and all firing was done by fuse. The quantity of explosive used averaged 7.5 pounds per cubic yard of heading rock. An attempt was made to reduce the quantity of explosive used, but it was found that smaller amounts would not break the rock into pieces of

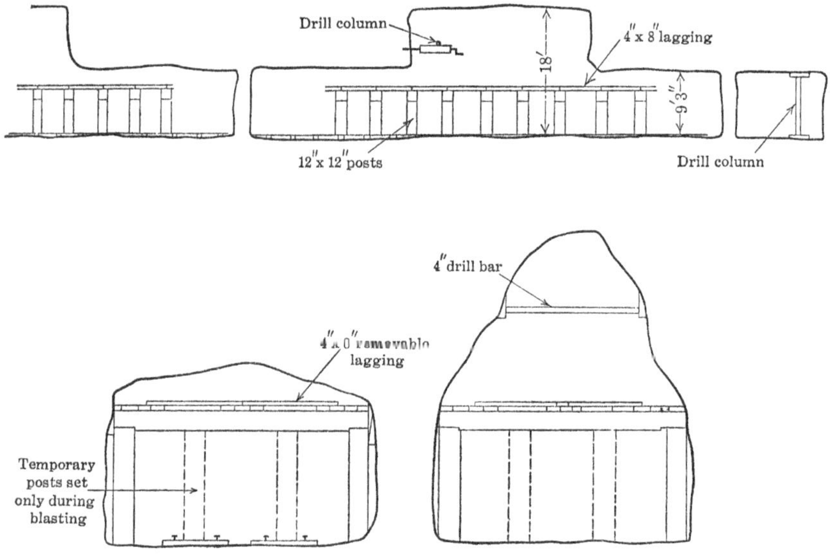

FIGURE 9.—Sections of Hunter Brook Tunnel, showing method of driving.

convenient size for handling. A round in the stope required 12 to 16 holes. The explosive consumption in the stopes was 2.5 pounds per cubic yard.

THE ELIZABETH TUNNEL.

SIZE AND CHARACTER OF TUNNEL.

The Elizabeth Tunnel[a] is part of the 217-mile aqueduct constructed by the city of Los Angeles, Cal. The methods used in

[a]Abstract from Aston, W. C., The Elizabeth Tunnel: Mines and Minerals, vol. 31, September, 1910, pp. 102–105.

driving are of particular importance because of the speed that was maintained. The tunnel was driven from both ends. It is 26,860 feet long and 12⅓ by 12¾ feet in cross section. When the work was first started the ground was soft and required timbering, but later harder rock was encountered, and, with the exception of a few softer belts at irregular intervals, it remained hard throughout. The rocks met were mainly granite, gneiss, or schist. A typical cross section of the tunnel, which ordinarily required 25 holes to round out, is shown in figure 10. The full heading was driven without benches, wings, or galleries.

METHOD OF DRILLING.

The bar a was first put in place and one machine on this bar drilled the 3 upper holes. Two machines used on a second bar, b,

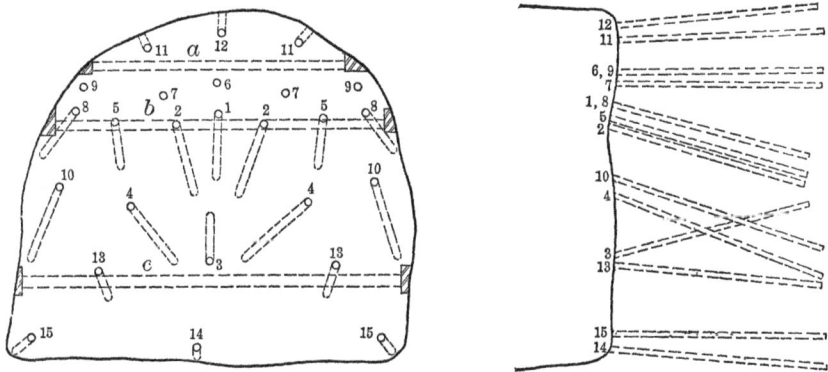

FIGURE 10.—Arrangement of drill holes in Elizabeth Tunnel.

drilled 15 holes. To drill the lower holes the second bar was lowered to a new position, or else a new bar, c, was placed in position and the machines from the second bar lowered to it as soon as the upper holes had been finished. The amount of explosives used varied from 15 to 40 pounds per foot of advance, depending on the rock encountered. Each shift was required to drill and blast, the length of the round being regulated by the nature of ground encountered in the first hole drilled. All holes were fired by fuse.

DATA ON TUNNEL DRIVING.

Data regarding this work during 1909 and the month of April, 1910, are given in the table following.

62626°—Bull. 80—15——6

Data on driving of south portal, Elizabeth Tunnel.

Item.	1909	April, 1910.
Total footage...	a 7,585	b 9,738
Footage...	4,476	604
Required progress, feet......................................	2,920	240
Average daily footage..	12.26	20.13
Required daily footage.......................................	8.00	8.00
Footage of untimbered section................................	3,274	604
Cost of untimbered section...................................	$121,787	$15,257
Footage of timbered section..................................	1,202	....
Cost of timbered section.....................................	$51,941	....
Average cost of timbering, per foot..........................	$1.16	....
Total bonus footage..	1,563	364
Total bonus pay roll...	$23,440	$2,377
Number of shifts worked......................................	1,055	90
Number of shifts lost..	40	....
Number of holes drilled......................................	21,066	1,924
Number of feet drilled.......................................	135,896	16,079
Total time of drilling, hours................................	3,656	324
Average time of drilling, minutes per foot of progress.......	49.01	....
Average time of drilling 1 foot of hole, minutes.............	1.61	1.209
Total amount of powder used (including trimming), pounds......	143,659	16,100
Average amount of powder used, pounds per foot...............	32.09	....

a End of 1909.   b End of April, 1910.

In the Jawbone division[a] of this same aqueduct very rapid progress was made, but the rock, which consisted of sand, gravel, cement, and decomposed granite, was much softer.[a] The nature of the ground permitted drilling by hand augers, broke up fine when shot, and could be easily shoveled. The roof was carefully arched and in this way the walls stood for weeks without timbering. This division was 9 feet 10 inches by 10 feet 8½ inches in section. During the month of August, 1910, the tunnel was driven 1,061 feet through dry cemented sand and sandstone. The work was done by hand and required 4,450 pounds of 40 per cent ammonia dynamite[b] and 2,372 pounds of black blasting powder.

LARAMIE-POUDRE TUNNEL.

SIZE AND CHARACTER OF TUNNEL.

The Laramie-Poudre Tunnel[c] is one of the largest irrigation tunnels in Colorado, being 12,000 feet long and 10 by 8 feet in section. The tunnel section is approximately an ellipse, it being considered more economical to drive than a rectangular section would be. The holes were drilled and shot in the manner and rotation shown in figure 11, and this required two horizontal set-ups of the tunnel bar.

[a] Herrick, R. L., Tunneling on Los Angeles aqueduct: Mines and Minerals, vol. 31, October, 1910, p. 136.

[b] Editorial, Explosives for tunnel driving: Mines and Minerals, vol. 31, October, 1910, p. 159.

[c] Abstract from article by Coy, B. G., The Laramie-Poudre tunnel: Eng. Rec., January 14, 1911, p. 45, and from article by Brunton, D. W., Notes on the Laramie-Poudre tunnel: Bull. Am. Inst. Min. Eng., April, 1912, p. 357.

The drill holes were 2⅛ inches in diameter at the collar and 1¾ inches at the point. Two and sometimes three water Leyner drills were used in drilling the holes, which varied from 8 to 10 feet in depth.

TESTS OF EFFICIENCY OF EXPLOSIVES.

In the early stages of the work numerous tests were made to determine the best and most economical explosive for use in the tunnel. Powders ranging from 100 per cent gelatin to 50 per cent dynamite were used in the work, the former particularly in loading the cut holes, but as a result of these experiments it was found that, except where the granite was extraordinarily hard and tough, 60 per cent gelatin dynamite gave as good results as explosives of higher grade.

In some of the 10-foot cut holes in which the 60 per cent gelatin dynamite failed to break out the rock satisfactorily good re-

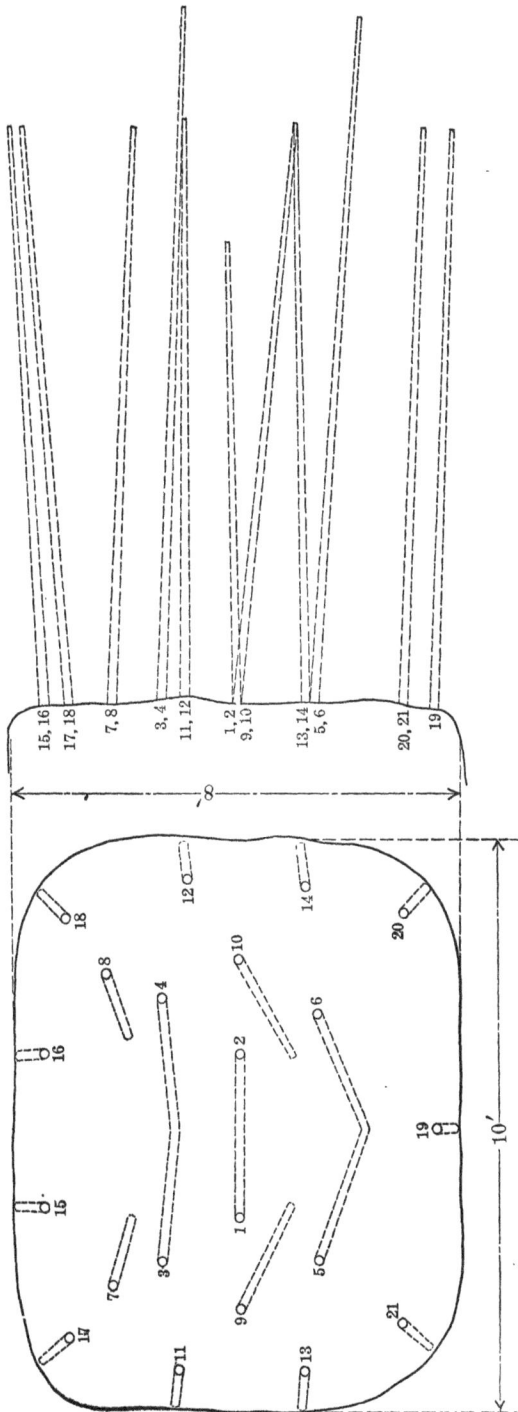

FIGURE 11.—Arrangement of drill holes in Laramie-Poudre Tunnel.

sults were obtained by using a blasting gelatin, containing 92 per cent of nitroglycerin for the bottom third of the charge and a 60 per cent gelatin dynamite for the remainder of the hole.

PRACTICE IN USE OF EXPLOSIVES.

The machine men loaded the holes, but the foreman superintended the loading and decided on the amount of gelatin dynamite to be used. A stick of 60 per cent gelatin dynamite was placed in the bottom of each hole and the primer above this. Five sticks were usually placed on top of the primer, except in the cut holes. They received 3 or 4 extra sticks. The three lifters were loaded to the collar, the additional amount of explosive used there being used for the purpose of throwing the muck as far back from the face as possible.

The usual practice of using stemming in the holes was discontinued early in the work, as it was believed that with heavy charges such as these the powder formed its own stemming, and that when the holes were loaded to the collar the rock was more thoroughly pulverized and, in consequence, more easily handled. The results of tests made at the Pittsburgh testing station of the bureau to determine the effect of stemming on the efficiency of an explosive of this type have shown that this assumption is wrong. A small quantity of moist clay stemming greatly increases the efficiency of all classes of explosives and, accordingly, if stemming had been used in this work the rock would have been more thoroughly pulverized than it was. It is believed by the authors that less explosives could have been used in this work and equally effective results would have been obtained if small quantities of moist clay stemming had been used in the collars of the drill holes.

It is true that in case of a misfire there is an advantage in having no stemming in a drill hole. Thus, in the work described, when there was a misfire an additional primer was placed against the unexploded charge and fired; it usually caused complete detonation of the charge.

USE OF FUSE.

Safety fuse having a rate of burning of about 1 foot in 40 seconds was used with No. 5 and No. 6 detonators (blasting caps) crimped to the fuse. The fuse was cut to exactly 10-foot lengths for all the holes, except for the lifters, for which the lengths were made 2 feet longer; the reason for the increased length of the lifter fuse was that as these holes could not be readily examined after the blast it was desirable to have them explode sometime after all the other shots so that the reports could be clearly heard and counted. In order to procure the proper sequence in firing, the foreman cut 20 inches from the ends of the fuse protruding from the short-cut holes, 18 inches from the fuse in the two uppercut holes, and so on.   Holes 1, 2, 3, 4, 5, and 6, comprising the cut, were spitted and shot first.   If the cut broke properly the remainder of the round was then fired.

Where the ground was wet the fuse was tarred for 1 foot back from the detonator (blasting cap), the tar being worked in around the edges of the detonator.   The remainder of the fuse, to within 2 feet of the end, was coated with heavy axle grease.

VENTILATION.

A Connersville blower, with a capacity of 13 cubic feet per revolution, and running at 225 revolutions per minute, was used for ventilation.   It supplied air through a 15-inch pipe laid to within 100 feet of the face.   It could clear the heading of gas in 20 to 30 minutes.

DRILLING RECORD.

The number of feet drilled each month on the tunnel at both portals is given below.   The rock was mainly a hard, tough granite.

Number of feet drilled per month at Laramie-Poudre Tunnel.

Month.	East portal.	West portal.	Month.	East portal.	West portal.
1910.	*Feet.*	*Feet.*	**1911.**	*Feet.*	*Feet.*
January	302	....	January	609	....
February	315	....	February	420	....
March	350	202	March	653	....
April	354	279	April	583	....
May	513	336	May	635	....
June	429	388	June	576	....
July	443	371	July	497	....
August	527	293	August	106	....
September	485	286			
October	420	28	Total, 1910 and 1911	9,123	2,183
November	424	....		2,183	....
December	482	....	Grand total	11,306	....

During March, April, and May, 1911, the record for distance driven, drilling, and powder consumed was as follows:

Distance driven, drilling progress, and powder consumed during three months in 1911, Laramie-Poudre Tunnel.

Month.	Length of tunnel completed.	Number of holes drilled.	Length of holes drilled.	Average length of holes drilled daily.	Sticks of powder used.
1911.	*Feet.*		*Linear feet.*	*Linear feet.*	
March	653	1,965	14,330	154	14,808
April	583	1,759	12,510	139	16,171
May	635	1,985	15,263	164	18,311

From these data, if it be assumed that the tunnel has a cross section of 8 square yards and that each stick of gelatin dynamite weighs 0.7 pound, 7 pounds of explosive was required for each cubic yard of rock excavated.

DATA ON HARD-ROCK TUNNELING IN AMERICAN TUNNELS.

This result is compared with the data set forth in the following table for other hard-rock tunnels:

Some records of American hard-rock tunneling.[a]

Name of tunnel.	Section.	Rock penetrated.	Best month's work.		Average for several months.		Amount of explosive used per cubic yard.
			Month.	Distance driven.	Number of months averaged.	Average.	
	Feet.			*Feet.*		*Feet.*	*Pounds.*
St. Pauls Pass, C. M. & P. S. Ry.	19 by 24	Quartz........	January, 1909.	385	12	272	[b] 5.4
Ouray (Colo.)............	7.5 by 7.5	....	January, 1908.	359	5	342	7.0
Los Angeles Aqueduct:							
No. 27................	9 by 10	Medium hard granite.	April, 1909.	464	4	348	3.2
Elizabeth.............	....	Gneiss, granite	March, 1909.	464	....	....	....
Kellogg (Idaho)	9 by 11	Quartzite.....	October, 1898.	354	....	....	....
Gunnison (Colo.).........	11 by 12	Granitic gneiss	January, 1900.	449	....	....	....
Rondout Siphon (Catskill).	c 17½	Hudson River shale.	November, 1909.	488	....	....	9.0
Chicago waterworks......	c 16	Fine-grained limestone.	April, 1909.	468	11	369	....
Buffalo waterworks......	12 by 15	Flinty limestone.	March, 1910.	390	6	354	3.0
Laramie-Poudre (Colo.) ..	8 by 10	Granite...	....	653	3	624	7.0
Hunter Brook (N. Y.) ...	15 by 18	Schist and gneiss.	....	283	5	273	5.0

a Engineering Record, Progress in hard-rock tunneling: Vol. 61, June 25, 1910, p. 800.
b Heading only.
c Diameter.

TRANS-ANDINE SUMMIT TUNNEL.

The Trans-Andine Summit Tunnel[a] is a railway tunnel on the line between Chile and Argentina. It is 9,933 feet long and in cross section is a replica of the Simplon Tunnel, the cross-sectional area inside the concrete being 272 square feet. A bottom heading having a sectional area of 107 square feet was driven from the Argentina side; 13 to 14 holes per round, each 6½ feet long, were drilled in the face. The excavation required 3.7 pounds of "gelignite" per cubic yard of rock. The maximum monthly progress was 272 feet.

From the Chilean side the tunnel was started by a top heading with a sectional area of 75 square feet. Fifteen to sixteen 5-foot holes were drilled per round, giving an average daily advance of 8.23

a Engineering Record, Tunneling methods in the Andes: Vol. 63, Feb. 18, 1911, p. 190.

feet against that of 6.75 for the Argentina side. On this side the excavation required 6.1 pounds of blasting gelatin per cubic yard of rock. The material penetrated was a reddish clay of varying hardness and a hard volcanic sandstone.

SIMPLON TUNNEL.

The Simplon Tunnel[a] is the longest tunnel in the world. It connects the towns of Brigue, Switzerland, and Iselle, Italy, a distance

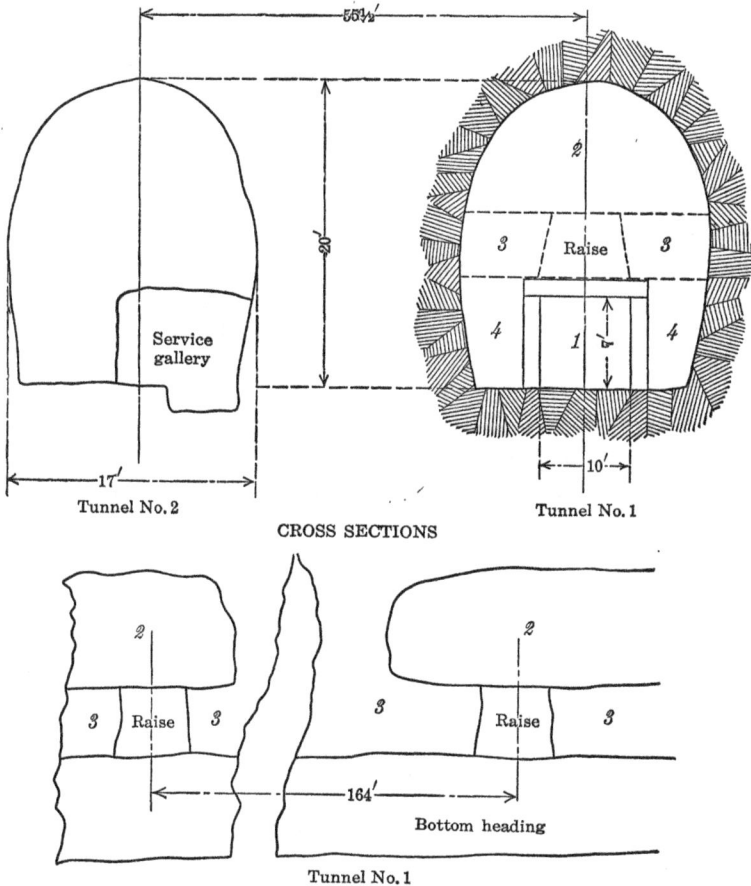

CROSS SECTIONS

Tunnel No. 1
LONGITUDINAL SECTION
FIGURE 12.—Sections of Simplon Tunnel.

of 12.4 miles. It consists of two parallel, singe-track, masonry-lined tunnels, spaced 55½ feet apart between center lines, and having a

[a] King, C. R., The construction of the Simplon Tunnel: Eng. News, vol. 50, 1903, pp. 174–179; Saunders, W. L., Tunnel driving in the Alps: Trans. Am. Inst. Min. Eng., vol. 42, 1911, pp. 441–446.

cross section of the shape and size shown in figure 12. At first only
Tunnel No. 1 was driven to its full cross section. A small 7 by 10
foot gallery was driven in Tunnel No. 2, mainly for the purpose of
ventilation and drainage.

In Tunnel No. 1 the center bottom heading *1* (fig. 12) was driven
first. This was also 7 by 10 feet in section, and was timbered
and covered with a closely boarded roof. From this heading a
raise was driven up to the roof line every 164 feet. The top heading
2 was then excavated by working in both directions from each raise.
The shallow transverse section *3* was then removed and finally the
two side "cheeks" *4, 4*. Rotary power drills mounted on carriages
were employed in driving the advance headings, the only part of the
work on which power drills were used.

With each round of holes the heading advanced 4½ feet, each blast
removing from 265 to 275 cubic feet of rock. The average rate of
advance was about 16 feet a day at the Italian end and 20 to 21 feet
at the Swiss end. The rock was mainly gneiss.

The explosives used were an "83 per cent" dynamite at the Italian
end and a "64 per cent" dynamite at the Swiss end. The explosives
were put up in packages of about 1 pound each, and each hole was
charged with 6 cartridges. Each round in the heading required 10
to 11 holes. An average of 6½ to 8½ pounds of dynamite was used for
each cubic yard of rock excavated.

The explosives magazine was at the Italian end, about 1½ miles
from the tunnel portal, and each day's supply of explosives was taken
from there and transported on a push car to a storehouse in one of
the cross galleries.

The charges were fired by means of ordinary fuse, which was cut
in different lengths to give the desired interval of time between the
firing of successive holes. Four to five rounds were fired daily. From
a report on three months' work on the Italian side in 1900 the following data were obtained:

Data on excavation of Simplon Tunnel.

Total length excavated with cross-sectional area of 62 square
 feet in 91 days_____feet__ 2,880
Number of holes drilled_____ 7,940
Total depth of holes_____feet__ 33,000
Dynamite used _____pounds__ 56,000
Dynamite used per cubic yard of rock_____do____ 8.5

A permanent plant, consisting of two fans 12.3 feet in diameter,
running at 400 revolutions per minute, was installed at each end, for
ventilating the tunnel during and after construction. The air pas-

sages from the ventilator house bifurcated near the tunnel ends and a door placed at the angle of the bifurcation permitted the closing of either fork of the passage. The tunnel portals were closed by sail-cloth curtains. Air could thus be circulated either by aspiration or compression in either tunnel as desired. At intervals of about 656 feet the main tunnel and service gallery were connected by transverse galleries, driven at an angle of 50°.

About 15 minutes was required after each blast to clear the heading of fumes. To accomplish this an exhaust ventilating pipe was run close to the face. A spray of water from the pressure pipes was also employed to absorb the gases and cool the air at the heading.

LOETSCHBERG TUNNEL.

The Loetschberg Tunnel [a] connects the cities of Berne and Brigue in Switzerland, and is one of the longest of the chain of Alpine

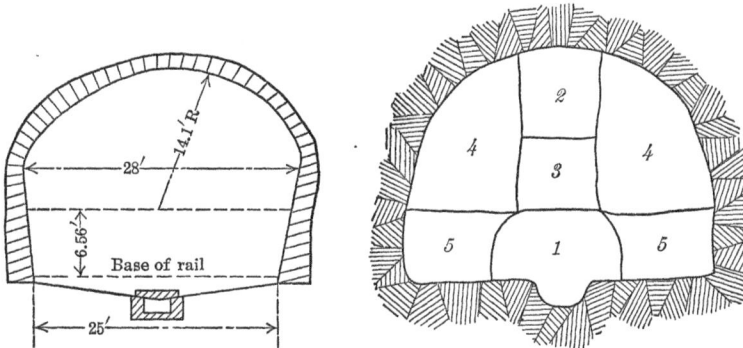

FIGURE 13.—Sections of Loetschberg Tunnel.

tunnels. The main tunnel is 47,678 feet long and is provided with double tracks. It was first planned to be on a tangent, but during the progress of the work a cave-in filled the tunnel for a length of 5,900 feet, and it was decided to insert a curve of 3,600 feet in the tunnel in order to drive through solid rock. In cross section the tunnel is of the type shown in figure 13. The width of the finished tunnel section is 28 feet at the arch springing and 25 feet at the base of the rail. The arch is semicircular, the crown being 20.7 feet above the base of the rail.

The sequence of excavation is shown in the excavation diagram. The bottom-heading method generally employed in Alpine tunnel practice was used. This heading was made 6½ by 10 feet in cross

[a] Saunders, W. L., Tunnel driving in the Alps: Bull. Am. Inst. Min. Eng. No. 55, July, 1911, p. 507.

section and was driven several hundred feet in advance of the enlargement. Upraises 500 to 600 feet apart were then driven to the roof, as in the Simplon Tunnel, and a top heading started back and forth. The top heading was then enlarged in the sequence shown in the diagram. When the inclination of the strata was nearly vertical, or the ground was of a treacherous nature, various modifications of the above method were employed. Compressed-air drills mounted on drill carriages were used for work in the heading. These carriages carried three to six drills mounted on a horizontal bar. The drill holes, having an average depth of about 4 feet, were started with a 3-inch drill and finished with a 2-inch drill. All holes were fired with fuses about 4 feet long, the center holes being fired first. The rate of drilling was 15 or 16 holes in about 1.1 hours. Drilling in the top heading was accomplished by air drills, but hammer hand drills were generally used for the enlargement.

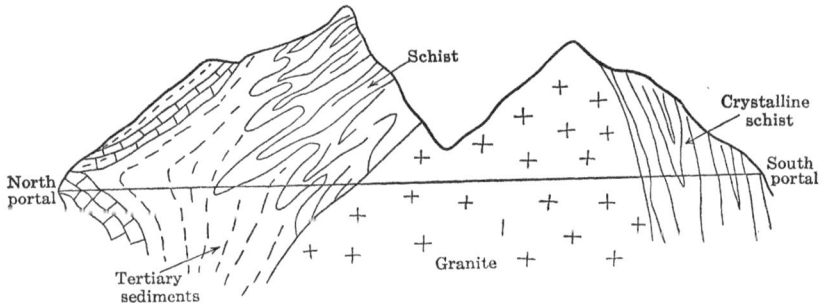

FIGURE 14.—Section showing rocks penetrated by the Loetschberg Tunnel.

Three kinds of explosives were used in the work. An "85 per cent" dynamite was used in the heading, and, with the shallow holes used, this broke the rock into small pieces and threw it back from the face so as to allow rapid clearing of the track. For enlarging the tunnel and for small blasts, Westphalite and Cheddite were used. The dynamite cartridges were wrapped with red paper in order to be easily detected in case of misfires. The dynamite carriers and handlers were also provided with red lanterns. As the workingmen were insured against accident or death by the contracting company, strict enforcement of the regulations was exercised in the handling and use of the explosives.

The rocks encountered in a section through the tunnel axis are shown in figure 14. On the north side there was calcareous material for a distance of 13,000 feet, on the south side crystalline schist was encountered for the same distance, and the center part was mainly granite.

When the headings approached each other great care was taken to prevent accident. The blasts on both sides were fired simultaneously at intervals of four hours, the watches of the engineers being compared daily by telephone. The usual length of fuse was increased to 11½ feet. The workmen retired 1,300 feet from the faces during the shots, and were retained 10 minutes after firing. The headings met March 31, 1911. Data relating to this work are shown in the following table:

Data on drilling and working conditions in driving north side (Kandersteg) and south side (Goppenstein) headings, Loetschberg Tunnel.a

[From official report of the Berner Alpine Railway.]

1908.

Item.	January, February, and March.		April, May, and June.		July, August, and September.		October, November, and December.		Total or average.	
	North side.	South side.	North side.	South side.	North side.	South side.	North side.	South side.	North side.	South side.
Progress of heading, feet	1,668	807	2,010	1,618	430	1,751		1,506	4,105	5,682
Average cross section of heading, square feet	64.52	62.40	63.48	68.88	63.48	63.48		64.52	63.60	64.52
Material broken out of heading, cubic yards	3,975	1,879	4,699	4,125	998	4,141		3,610	9,672	13,755
Number of working days	84	89	83	89	84	90½		88½		
Average daily progress, feet	18.85	14.79	25.78	18.30	18.66	19.34		16.92	21.09	17.33
Average progress of one round, feet	3.83	3.91	4.19	3.54	3.83	3.44		3.18	3.95	3.51
Number of rounds	457	206	505	459	112	510		473	1,074	1,648
Drilling time of one round, hours	1.40	2.20	1.17	1.35	1.23	1.15		1.29	1.27	1.40
Mucking time of one round, hours	2.45	3.20	1.37	2.53	3.10	2.55		2.51	2.30	3.00
Entire duration of one round, hours	4.35	6.22	3.58	4.40	4.34	4.15		4.31	4.22	4.58
Number of bore holes per round	12.5	11.4	12.8	12.0	12.5	11.7		12.0	12.6	11.8
Average length of holes, feet	4.13	4.82	4.46	4.62	4.43	3.54		4.33	4.33	4.33
Length of holes per cubic yard, feet	4.75	5.82	6.15	6.18	6.25	5.04		6.63	5.72	5.92
Dynamite per cubic yard, pounds	6.60	5.66	6.38	6.05	5.88	5.94		7.05	6.28	6.17
Number of steel bits per cubic yard	2.57	1.27	1.86	1.34	1.45	2.85		3.57	1.96	2.26
Number of drill carriages in operation	3.2	3.2	3.4	4	3	4		3.2	3.2	3.8
Air pressure at compressor, pounds			113	99.3	110.8	95	113	90.8	113	99.3
Air pressure at face, pounds			106.5	85.2	106.5	72.6	106.5	80.6	99.3	78.1
Air temperature at face, °F	78	85.2	55.4	75.4	51.4	76.6		78.8	53.2	76
Rock temperature at face, °F	53	71.6						(h)		
Remarks	(b)	(c)	(d)	(e)	(f)		(g)	(h)		

1909.

Item.	January, February, and March.		April, May, and June.		July, August, and September.		October, November, and December.		Total or average.	
	North side.	South side.	North side.	South side.	North side.	South side.	North side.	South side.	North side.	South side.
Progress of heading, feet	937	1,402	2,775	1,402	2,848	1,562	985	1,422	7,550	5,772
Average cross section of heading, square feet	70.15	69.90	71.00	66.7	67.0	67.2	83.9	66.7	73.0	67.6
Material broken out of heading, cubic yards	2,437	3,550	7,285	3,460	7,080	3,890	3,047	3,510	19,849	14,410
Number of working days	36½	90	85½	87	90	90	53½	88½	259½	355½
Average daily progress, feet	25.7	15.5	32.3	16.2	33.9	17.3	18.5	16.0	27.6	16.3
Average progress of 1 round, feet	4.23	3.18	4.82	3.41	4.85	3.70	4.33	4.16	4.55	3.62
Number of rounds	221	440	573	408	588	420	227	342	1,609	1,670
Drilling time of 1 round, hours	1.16	1.35	1.05	2.18	1.03	2.22	1.14	2.54	1.12	2.17
Mucking time of 1 round, hours	2.35	3.04	2.26	2.40	2.20	2.37	2.50	3.09	2.33	2.53

Entire duration of 1 round, hours			3.35	5.07	3.26	5.08	5.39	6.13	4.08	5.18
Number of bore holes per round	14.5	12.7	14.24	13.72	14.49	12.77	14.1	14.0	14.3	13.3
Average length of holes, feet	4.72	4.16	5.01	4.26	5.05	4.36	4.72	4.52	5.74	4.32
Length of holes per cubic yard, feet	6.25	6.60	5.64	6.91	6.11	6.08	4.99	6.20	5.48	6.44
Dynamite per cubic yard, pounds	6.57	7.64	5.53	8.19	0.65	7.36	4.21	6.79	0.72	7.48
Number of steel bits per cubic yard	0.85	3.66	0.72	5.42	4.0	6.59	0.66	4.0	4.0	5.38
Number of drill carriages in operation	3.98	4.0	4.0	5.1		4.9	4.0			4.5
Air pressure at compressor, pounds	113.6	78	106.5	79.5	106.5	88	106.5	110.8	108.2	89.5
Air pressure at face, pounds	106.5	63.9	99.4	71	97.9	65.3	99.4	63.5	100.8	65.9
Air temperature at face, °F	50.3	80.5	60.4	81.5	63.2	82.4	59.5	82.7	58.3	81.8
Rock temperature at face, °F	48.2	78.8	56.8	81.7	59.4	86.0	56.1	86.5	55.1	83.2
Remarks	(d)	(f)	(k)	(i)	(d)	(m)	(m)	(o)	(r)	

1910.

Progress of heading, feet	2,250	1,450	2,490	1,500	4,740	2,950
Average cross section of heading, square feet	70.8	66.5	67.6	63.5	69.2	64.6
Material broken out of heading, cubic feet	5,890	3,570	6,230	3,530	12,100	7,100
Number of working days	86½	84¼	86	90	172½	174½
Average daily progress, feet	26.0	17.15	28.9	16.65	27.4	16.9
Average progress of 1 round, feet	4.13	3.97	4.07	4.2	4.1	4.07
Number of rounds	541	366	613	358	1,157	724
Drilling time of 1 round, hours	1.18	2.31	1.13	2.48	1.15	2.40
Mucking time of 1 round, hours	2.19	2.50	2.02	3.10	7.10	3.00
Entire duration of 1 round, hours	3.50	5.32	3.22	6.02	3.36	5.47
Number of bore holes per round	14.1	14.0	14.89	14.3	14.5	14.15
Average length of holes, feet	4.76	4.43	4.56	4.69	4.66	4.56
Length of holes per cubic yard, feet	6.24	6.49	6.71	6.84	6.53	6.62
Dynamite per cubic yard, pounds	6.17	7.09	6.94	6.88	6.53	6.98
Number of steel bits per cubic yard	2.83	6.42	4.32	6.83	3.57	6.62
Number of drill carriages in operation	4	4	4	4	4	
Air pressure at compressor, pounds	121	96	121	99	121	98
Air pressure at face, pounds	106	71	114	92	110	81
Air temperature at face, °F	59.4	83.7	66.2	86.6	62.8	85.2
Rock temperature at face, °F	60.1	90.3	61.7	97.3	60.8	91.4
Remarks	(p)	(g)	(r)	(g)	(r)	(q)

a Saunders, W. L., Tunnel driving in the Alps; Trans. Am. Inst. Min. Eng., vol. 42, 1911, pp. 467–468.
b March 8 to 10, work stopped; rock conditions extremely favorable; mountain limestone.
c February 20 to March 31, work stopped; avalanche accident.
d Mountain limestone.
e Ventilation only one-half that on north side; chlorite and sericite gneiss.
f July 24, cave-in at Kander; 25 men buried.
g Tunnel driving stopped.
h Gneiss.
i This figure works out 29.1 when dividing total progress by number of days.

j Dolomite, slate, etc.
k Black mountain limestone.
l Various sedimentary limestones, slate, and granite.
m Granite, laccolithic. Eastern granite.
n Malm Trias.
o Porphyry.
p Granite.
q Granite and quartz porphyry.
r Eastern granite.

NOTE.—On the south side, after blasting, 12 connection valves were opened for the rapid removal of the air in the tunnel. During drilling compressed air was also used for ventilation.

MAGAZINES AND THAW HOUSES.

STORAGE AND HANDLING OF EXPLOSIVES.

In order that the demands of the users of explosives may be promptly supplied, it is necessary that manufacturers, transporters, distributors, and users of explosives maintain proper and adequate storage facilities. The necessity therefore arises of having magazines in which large quantities of explosives may be stored. At high temperatures explosives may become unstable and acquire an increased sensitiveness to explosion from shock or frictional impact; consequently they should be stored in cool places. As the addition of moisture damages explosives, they should never be stored in a damp place. As any change in composition may affect the safety or the efficiency of an explosive, explosives containing a definite quantity of moisture should not be so stored as to lose any of their moisture by drying out. Hence, all explosives should be stored in cool, thoroughly ventilated magazines situated on well-drained ground.

PLACING A MAGAZINE.

In choosing a site for a magazine the relation of the building to the ground level may affect two essential safety factors in opposite ways. When the magazine is placed above ground, adequate drainage is easily obtained and maintained, but any barricade, natural or artificial, required for limiting the effect of possible explosions is then less effective. When the magazine is placed partly or wholly below the surface of the ground, such barricades are more effective, and for some magazines the earth itself would be an adequate barricade, but such an arrangement may make proper drainage difficult and even impossible, and drainage should be given first consideration.

A certain magazine in eastern Pennsylvania is so placed in a hole that the top of the magazine is below the ground level, but as the ground slopes from the magazine, proper drainage is easily obtained by means of a large pipe leading from the base of the magazine. Entrance to the magazine is effected by means of steps leading down from the ground level to the drainage level.

94

PROTECTION OF EXPLOSIVES IN MAGAZINES.

During storage, explosives should be protected as far as practicable against heat, moisture, fire, lightning, projectiles, and theft. The buildings should therefore be weatherproof, covered by or made from fireproof and bullet-proof material, well ventilated, placed a long distance from other buildings, and not exposed to fire risk from burning grass or underbrush or combustible material about them.

Detonators or other devices containing fulminating composition should not be kept in a magazine in which there are explosives. For storing such devices a special small building erected at a safe distance from other magazines should be provided.

RECEIVING EXPLOSIVES.

A trustworthy person should be designated as a magazine storekeeper and should have sole supervision of the receiving and storing of explosives in magazines. He alone should have access to the magazines for the purpose of taking out explosives that are required for immediate use.

When a new consignment of explosives is received, it should be stored in the magazine in such a way that the oldest explosives will be issued first. When the packages are stored in a magazine a space of a few inches should be left between the walls and the packages for ventilation. The boxes should be so placed that the cartridges do not stand on their ends, because this position increases the rate of exudation or leakage of nitroglycerin from cartridges of nitroglycerin explosives.

OPENING PACKAGES.

Packages of explosives should never be opened within the magazine, but in a properly sheltered place at a safe distance from the magazine. They should be opened with a wooden mallet and a wooden wedge. If a screw driver is necessary in opening boxes, it should be used only for removing screws. Packages of explosives should never be opened with a nail puller, nor should powder kegs be opened with a pick.

REPAIR AND CARE OF MAGAZINES.

When it is necessary to repair or make alterations in a magazine all explosives should be carefully removed and the magazine thoroughly washed. All tools used in making repairs should be of wood or brass. The driving of nails into the floors of buildings that have been used in the manufacture of explosives has resulted in serious accidents.

Magazines should be kept clean and the floors should be kept free from grit or dirt. The ground around the magazine should be kept free from rubbish, leaves, and dead grass, and from other materials that might feed a fire. A warning somewhat like the following should be conspicuously posted about the magazine: "Explosives—dangerous—no shooting allowed."

No artificial heat of any kind, for thawing or other purposes, should be introduced into a magazine, for, as stated elsewhere, explosives may become unstable and are always more sensitive to shock when stored at high temperatures. When thawing is necessary the explosives necessary for immediate use should be taken away from the magazine and thawed in the manner described hereafter.

BULLET-PROOF MAGAZINES.

In the United States, especially in the Western States, where the hunting of wild game is general, bullet-proof magazines are essential. Accordingly the light-construction type can not be recommended.

In an investigation of the most suitable materials for magazine construction,[a] investigators of the Bureau of Mines tested fireproof materials in a finely divided state, to determine the resistance such materials may offer to the penetration of rifle bullets. It was assumed that the materials could be used as a filling between thin boards, which would act as a form and remain as part of the permanent structure. Sand has been used in constructions of this kind, but it is objectionable in that it eventually flows out through the cracks of the form to the floor of the magazine. Any gritty material of this kind on the floor is obviously a menace to safety. Mineral wool has been suggested for use as a suitable bullet-resisting material, but it has been found to have little value for this purpose.

LIGHTNING CONDUCTORS.

In England there is a statutory requirement that all magazines shall be provided with a lightning conductor, but no definite arrangement of the conductors is prescribed. In other countries a metallic network is suspended above the magazine on metal poles, the poles being placed in a circle around the magazine and 20 to 30 feet distant. At some magazines the network is not used, the poles alone being depended on for protection against lightning. A practice in Germany is to erect vertical copper rods of high electrical capacity near the magazine. The rods are 36 to 80 feet high, and are sometimes placed on the embankment.

USE OF GALVANIZED-IRON ROOF COVERING.

The magazine of the Bureau of Mines (Pl. XV, A, and fig. 15) has its entire roof and all sides covered with galvanized sheet iron,

---

[a] Hall, Clarence, and Howell, S. P., Magazine and thaw houses for explosives: Technical Paper 18, Bureau of Mines, 1912, 34 pp., 1 pl., 5 figs.

A. BUREAU OF MINES CEMENT-MORTAR MAGAZINE.

B. BUREAU OF MINES CEMENT-MORTAR THAW HOUSE

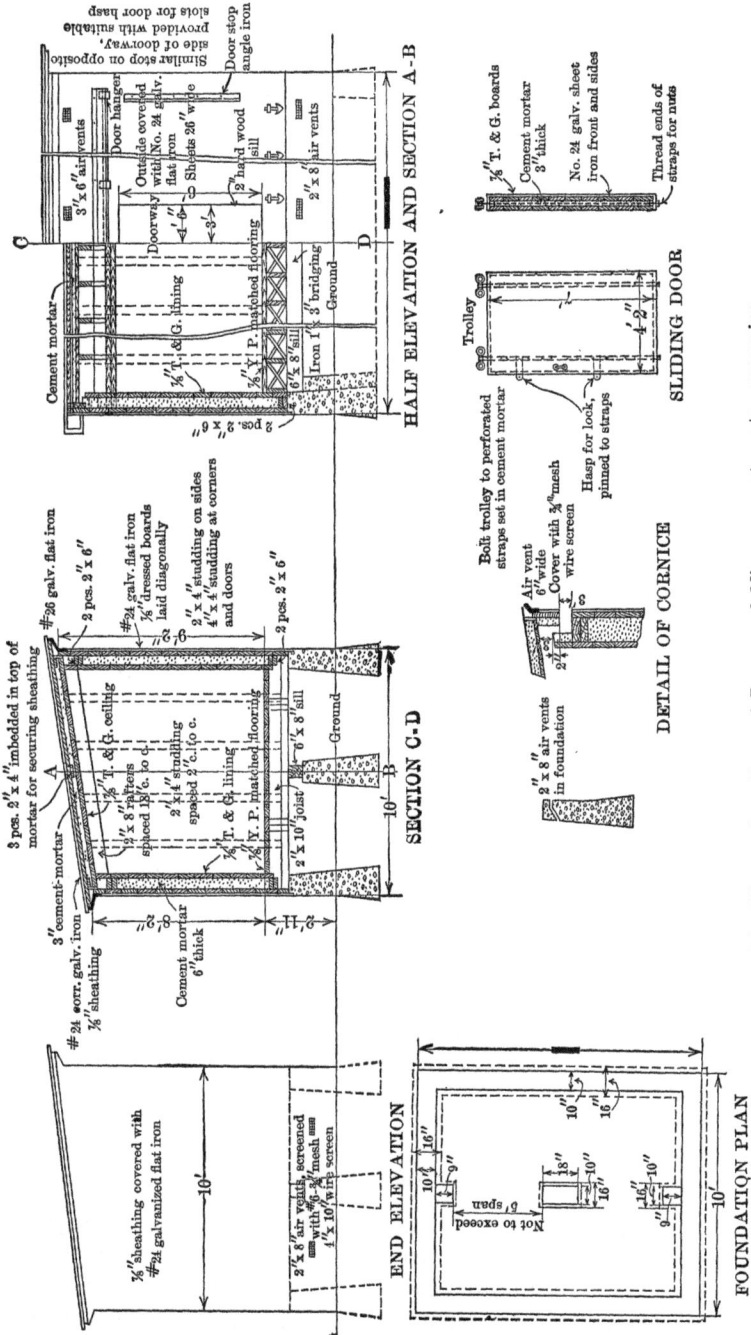

FIGURE 15.—Plan and sections of Bureau of Mines cement-mortar magazine.

HALF ELEVATION AND SECTION A-B

SLIDING DOOR

DETAIL OF CORNICE

SECTION C-D

END ELEVATION

FOUNDATION PLAN

and the iron parts are grounded with ½-inch iron rods extending from the four lower corners of the building, the ends of the rods being buried below the moisture line and being attached to a properly laid earth plate. The rods serve as excellent mediums for silently reducing the potential between the earth and the clouds. Where water mains or running streams are available, good practice requires that the "grounds" be properly attached to or immersed in them. If, as the result of an unusual electric discharge, any part of the galvanized-iron cover should be burned or melted, the explosives may still be protected from fire by the intervening 6-inch walls which are constructed of nonconducting cement mortar.

PROTECTION OF LIFE AND ADJACENT PROPERTY.

The most important result to be achieved in the storage of explosives is the prevention of explosions. If this is accomplished no damage to life and property can result. As absolute prevention is impossible, certain precautions must be taken to limit the damage should an explosion occur. These are discussed below.

SELECTION OF MAGAZINE SITE.

In selecting a site for a magazine the topography of the ground should be carefully considered, and advantage should be taken of any natural protection offered by hills. It is preferable to erect a magazine on sandy soil rather than on rocky ground, for, in the event of an explosion, the distance through which the earth waves are transmitted is materially reduced when the magazine rests on loose and friable ground. The distance separating magazines from inhabited dwellings and railways should be such that if an accidental explosion occurs within a magazine the least possible damage will be done to the buildings of the surrounding country. The laws in foreign countries require that between magazines and inhabited dwellings there shall be certain distances, depending on the quantity of explosives stored, the natural or artificial protection provided, and the position of the inhabited dwellings of the neighborhood.

COMBINED TABLE OF DISTANCES.

A combined table of distances fixed by different foreign and American authorities and showing for certain quantities of explosives stored the minimum lawful distance, where known, that a magazine may be located from inhabited structures, is given below. The proposed American distances are made to apply to nonbarricaded magazines, in order that a comparison with the distances of other countries may be readily made. The Austrian, Italian, and

Prussian distances in the table seemingly do not discriminate as to the allowable distances at which widely different quantities of explosives may be stored, but this seeming oversight is a result of the particular quantities used in the illustration, and the fact that any given allowable distance is applicable to a larger range of quantity.

Combined table of distances for the isolation of magazines from inhabited structures.

Quantity of explosives stored.	Barricaded magazine.	Nonbarricaded magazine.					
	Proposed American distances.	Proposed American distances.	Massachusetts distances.a	Austrian distances.	English distances.	Italian distances.	Prussian distances.
Pounds.	*Feet.*	*Feet.*	*Feet.*	*Feet.*	*Feet.*	*Feet.*	*Feet.*
50	120	240	270	328	....	....	330
100	180	360	351	328	....	....	330
500	400	800	606	1,640	300	230	495
1,000	530	1,060	861	1,640	450	230	495
5,000	780	1,560	1,716	3,280	960	574	999
10,000	890	1,780	2,166	3,280	1,575	656	999
50,000	1,460	2,920	5,550	....	5,550	1,640	....
100,000	1,835	3,670	10,500	....	10,500	1,640	....
500,000	2,755	5,510	....	....	....	....	....
1,000,000	3,455	6,910	....	....	....	....	....

a The State of Massachusetts is one of the few States that have enacted regulations governing the isolation of magazines.

BARRICADES.

In nearly all foreign countries mounds or artificial barricades are used about magazines having no natural protection. The mounds are constructed of earth or sand, and care is taken that large stones or any material that might be projected in large masses in the event of an explosion shall not be used in them. The mounds are usually 3 feet higher than the highest point of the roof. They are usually 3 feet wide at the top, with a natural slope on both sides, the inner slope reaching to within a few feet of the sides of the building. When such mounds are used the distances required between magazines and dwellings are materially less than when they are not used.

When a large quantity of explosives is stored barricades should always be erected if there is not a sufficient natural protection. Any side or sides of a magazine in line with any dwelling, shaft, highway, or railway should be guarded with mounds of more substantial construction than guard the other sides.

CONSTRUCTION OF MAGAZINES.

In existing practice the different magazines differ materially in construction, many of them being substantial structures built of concrete, brick, or large blocks of stone. These materials are objectionable in that in the event of an explosion within a magazine parts of the structure may be thrown as missiles. More frequently magazines are built of a light framework of wood covered with corrugated iron, no attempt being made to protect the explosives stored in them from the penetration of rifle bullets.

BUREAU OF MINES MAGAZINE.

As a result of experiments and of information furnished by manufacturers of explosives, a cement-mortar magazine (Pl. XV, *A*, and fig. 15) has been erected by the Bureau of Mines. The magazine has a capacity of 20,000 to 30,000 pounds of explosives, and was built at a cost of $400. The outside dimensions are 10 by 14 feet. The salient features of the magazines are its cement-mortar walls, sliding door, and roof. The cement mortar is 6 inches thick in all walls and 3 inches thick in the roof and the door. The door is secured by two substantial locks. No metal of any kind is exposed on the inside of the magazine. The ventilators above the floor are so arranged as to prevent the entrance of bullets or firebrands.

The means provided for ventilation have been found to be adequate, and accordingly, the storage conditions are favorable for keeping the explosives from deteriorating. The cement-mortar construction is effective in resisting the penetration of rifle bullets, and because of its friable nature it offers an additional advantage for the reason that, in the event of an explosion in or near the magazine, large masses of material will not be projected about the surrounding neighborhood. The galvanized-iron covering is fire resisting; it also serves as an excellent medium for protection against lightning, as the four corners of the building are properly grounded with metal rods, as above described.

In order that the details of construction may be thoroughly understood, the following bill of materials is presented. The concrete of the foundation consists of one part cement, three parts sand, and five parts gravel. The cement mortar in the walls, door, and roof consists of one part cement to six parts coarse sand.

BILL OF MATERIALS.

Lumber.

1 piece of yellow pine 2 inches by 6 inches by 18 feet.
1 piece of yellow pine 6 inches by 8 inches by 14 feet.
11 pieces of yellow pine 2 inches by 10 inches by 10 feet.
2 pieces of yellow pine 1 inch by 3 inches by 12 feet.
4 pieces of yellow pine 2 inches by 10 inches by 10 feet.
8 pieces of yellow pine 2 inches by 6 inches by 14 feet.
9 pieces of yellow pine 2 inches by 8 inches by 10 feet.
6 pieces of yellow pine 2 inches by 4 inches by 7 feet.

6 pieces of yellow pine 2 inches by 4 inches by 8 feet.
2 pieces of yellow pine 4 inches by 4 inches by 7 feet.
2 pieces of yellow pine 4 inches by 4 inches by 8 feet.
24 pieces of yellow pine 2 inches by 4 inches by 9 feet.
3 pieces of yellow pine 2 inches by 4 inches by 14 feet.
9 pieces of hemlock 1 inch by 10 inches by 10 feet.
8 pieces of hemlock 1 inch by 10 inches by 12 feet.
15 pieces of hemlock 1 inch by 10 inches by 14 feet.
28 pieces of hemlock 1 inch by 10 inches by 16 feet.
400 feet of lumber for framework of foundation.
48 board feet of 16-foot No. 2 white-pine flooring.
856 board feet of No. 2 yellow-pine flooring consisting of 19 bundles 16 feet
long, 1 bundle 14 feet long, and 11 bundles 10 feet long.

Hardware.

100 pounds of 8d. wire nails.
40 pounds of 20d. wire nails.
4 pieces of iron $\frac{3}{8}$ inch by $1\frac{1}{4}$ inches by 7 feet.
2 pieces of iron $\frac{3}{8}$ inch by $1\frac{1}{4}$ inches by 18 inches.
2 pieces of angle iron 3 inches by 5 feet.
2 pairs of No. 130 Coburn trolley hangers.
8 feet of No. 4 track.
2 end brackets.
1 center bracket.
1 heavy iron door handle.
4 brackets with hardwood rollers.
500 square feet of No. 26 gage galvanized flat iron.
150 square feet of galvanized corrugated iron.
2 pieces of 5-inch by 8-inch, $\frac{3}{4}$-inch mesh, No. 6 wire screen.
8 pieces of 4-inch by 10-inch, $\frac{3}{4}$-inch mesh, No. 6 wire screen.

Cement, Sand, and Gravel.

80 bags of cement.
200 bushels of sand.
150 bushels of gravel.

THAWING EXPLOSIVES.

When exposed to a temperature of 52° F. or less for a relatively short time, the nitroglycerin in ordinary dynamites crystallizes, or, as is commonly stated, freezes. In order to use such explosives safely and to procure the maximum efficiency they must be properly thawed before being used. While being thawed, the explosives should be kept in a horizontal position, as otherwise they are more liable to permit the nitroglycerin contained in them to exude.

PRECAUTIONS IN THAWING.

The thawing of frozen explosives requires extreme care, and the use of improper methods has led to serious accidents. No attempt should be made to thaw a frozen explosive by placing cartridges of it before a fire, or near a boiler, or on steam pipes, or in hot water, or in the sunlight. While being thawed, nitroglycerin explosives are extremely sensitive and should be handled with great care. During thawing the nitroglycerin tends to separate from the " dope " and to run out of the cartridge (that is, to exude), and this is a source of danger.

THAWING SMALL QUANTITIES OF EXPLOSIVES.

When only a small quantity of the explosive is required, it may be thawed in a thawer such as is furnished by manufacturers of explosives, a device that has been found safe for use as directed. Such a thawer is shown at the left of the foreground in Plate XV, *B*. It consists of a water-jacketed tin vessel, closed with a tin cover, in which the cartridges are placed. Before the water is placed in the vessel it is warmed to a temperature not uncomfortable to the immersed hand; the temperature should never exceed 130° F. The cartridges are allowed to remain in the thawer until gentle pressure shows that they are completely thawed throughout. When thawed, the material feels plastic, or like a mass of flour, between the fingers. When frozen, or partly frozen, it feels somewhat rigid and hard. The stick should be thawed completely, because dynamite when frozen can be detonated only with great difficulty, and any part that is frozen will be imperfectly detonated in the blasting hole; hence not only may such partly frozen powder fail to give its full effect as an explosive, but there is danger of a serious accident in a coal mine where the explosive is used, as it may give rise to a "blown-out" shot; if a "blown-out" shot results, the burning solid part of the dynamite or its flame may set fire to combustible dust or fire damp in the air of the mine, and will yield poisonous gas.

THAWING BY MANURE.

When explosives are used in temporary projects in quantities that do not require the thawing of more than 200 pounds at one time, use is often made of manure-thawing boxes. They are simple in construction, consisting of tight boxes, which, after the explosives have been inserted, are completely surrounded with 12 to 18 inches of fresh manure. Such thawing boxes have the advantage of cheapness, but they are adapted to the thawing of explosives only when the explosives are not needed promptly, because they are slow in action. The somewhat common practice of placing cartridges of explosives directly in manure piles can not be recommended, for then the cartridges may absorb moisture and, moreover, manure may, through fermentation, become as hot as 150° F., a temperature that is unsafe for use in the thawing of any explosive.

THAWING LARGE QUANTITIES OF EXPLOSIVES.

Where large quantities of explosives are used frequently, a thaw house situated at a safe distance from the magazines and other buildings should be provided. The capacity of the thaw house should be so limited that it will not hold more than is necessary for one shift. It should never hold more than 500 pounds of explosives. The best practice is to place in the thaw house only the amount of explosive that is required for immediate use. It is not a safe practice to expose explosives to long-continuous heating, as some explosives, such as gelatin dynamites, have been known to decompose and explode when subjected to high temperatures for but a few days.

PLACING OF THAW HOUSES OR MAGAZINES IN COLD CLIMATES.

In several of the metal mines in the northern part of this country, where extremely low temperatures prevail in winter, it has been found necessary to erect thaw houses within the mines. The thaw houses are usually placed in an abandoned or worked-out place in the mine in one of the upper levels. Thaw houses so located have been heated by electricity, a cluster of incandescent globes or a series of resistance coils being used. Many such thaw houses have the source of heat directly under the explosives, and many accidents have resulted as a consequence of this arrangement. Other thaw houses situated in mines have had a thin sheet of galvanized iron interposed between the source of heat and the explosives in order to prevent the explosives from coming in contact with the electric coils, but nearly all of them lack the proper means of control by which to keep the temperature under a safe limit, with the result that the explosives sometimes get so hot as to become dangerous.

TRANSPORTING THAWED EXPLOSIVES.

When an explosive is used in mines or quarries situated in extremely cold localities it should be transported from the thaw house to the working place in an insulated container of the required size, which is provided with a jacket of a good nonconducting material, such as hair felt. A device of this kind prevents the thawed explosives from freezing during transportation, and, accordingly, it insures the greatest efficiency in their use and permits the erection of the thaw houses outside of the mine.

MANURE THAW HOUSES.

Manure thaw houses have been satisfactorily used for explosives employed in temporary projects in quantities that require the thawing of more than 200 pounds at one time. The principle of their construction is similar to that of manure thawing boxes, except that the cartridges of explosives are placed in them in layers on shelves. This arrangement compensates for the reduced area of radiation per unit of volume of the inner compartment. In manure thaw houses a door must be provided for entering the compartment, and the roof and lower part of the exterior retaining walls should be so constructed that they are easily removable, in order that fresh manure may be added when necessary.

As the heat generated by manure is produced by biochemical reactions, manure obviously will be effective as a thawing material only as long as the reactions are efficiently maintained. Such heat generation often continues for many weeks. The possible high temperature within the manure walls of the thaw house or the thawing box .are not considered dangerous, because the explosive within the compartment is isolated from the manure. Although careful search has been made, no instance has been found of fire resulting from the natural heat of manure used in thaw houses or thaw boxes in the manner just described.

CONSTRUCTION OF THAW HOUSES.

What has been said of the construction of magazines applies equally to the designing of thaw houses. In addition to protecting thaw houses from dangers without, the increased sensitiveness of high explosives, owing to the exposure of the explosives to the high temperatures that must be maintained within the thaw house, constitutes an additional danger.

Thaw houses are usually so situated as to be convenient to the shaft, pit mouth, or place where the explosives are to be used, and consequently they should be constructed of such material as would not be projected in large pieces should an accidental explosion occur within them. It is also essential that thaw houses be bullet proof and protected from lightning, unlawful entry, and fire.

In constructing thaw houses some source of heat that can be kept within safe limits must be introduced. Low-pressure steam is usually available, and when used in the manner hereafter described at pressures not exceeding 3 pounds it is effective and one of the safest means of thawing explosives. The temperature of the air entering a compartment should never exceed 130° F., and a lower temperature is desirable if a temperature of 90° F. can be maintained within the thaw house with the steam at the lower temperature.

BUREAU OF MINES THAW HOUSE.

The cement-mortar house (Pl. XV, B, and figs. 16 and 17) recently built by the Bureau of Mines for thawing large quantities of explosives has a capacity of 500 pounds and cost, complete, $200. It was erected after a consideration of the various details involved, is not of prohibitive cost, and offers the following advantages: The cement-mortar walls, roof, and doors protect the explosives from bullets; the cement-mortar and galvanized-iron covering are fire-resisting, and the latter furnishes a good conductor for lightning; the building is substantial; entrance is difficult and is unnecessary in practical use, because the trays within it can be entirely removed without anyone entering the house (see Pl. XV, *B*); all parts of the house are accessible for cleaning; the explosives can be distributed in thin layers in the trays; the thermometers are easily read from without.

This thaw house also offers the following advantages in regard to temperature: No high temperature from artificial sources of heat occurs in the vicinity of the thaw house; the low-pressure steam, hot-water, or electric coils used are not in the thaw house proper, but in a separate compartment at its side, and hence particles of explosive can not come in contact with them; the explosives are not subjected to the evil effects of free steam or water; the temperature is surely and easily controlled with little attention; the entrance of cool air or the escape of heat from the thaw house is reduced to a minimum; the stack used makes possible the positive circulation of heated air. The position of the heating box with reference to the explosives compartments, the slanting roof of the heating box, and the stack make a natural draft easy.

Its operation is as follows: The cold air from without is drawn through the air regulator, over the hot coils, through the damper opening, and into the air conduit. Thence it is passed upward through the holes in the compartment regulators and through them only. The air is then deflected to each side of each compartment by means of deflector boards, one of which is directly above each compartment regulator. As the trays in each compartment are staggered, the air is deflected backward and forward as it rises, until it finally escapes through the stack.

The air regulator consists of five adjustable three-quarter inch louver boards spaced 3 inches apart. The regulator is used only to control the quantity of air passing over the coils. If the outside air is very cold, the quantity of air is reduced by partly closing the louver boards, in order that the air entering the explosives compartments may be heated to the proper temperature. On the other hand, when the outside temperature is comparatively high, say 32° F., the louver boards are left completely open, or nearly so, in order that the air entering the explosives compartments may not become too warm for the safe thawing of explosives. A damper is included in the con-

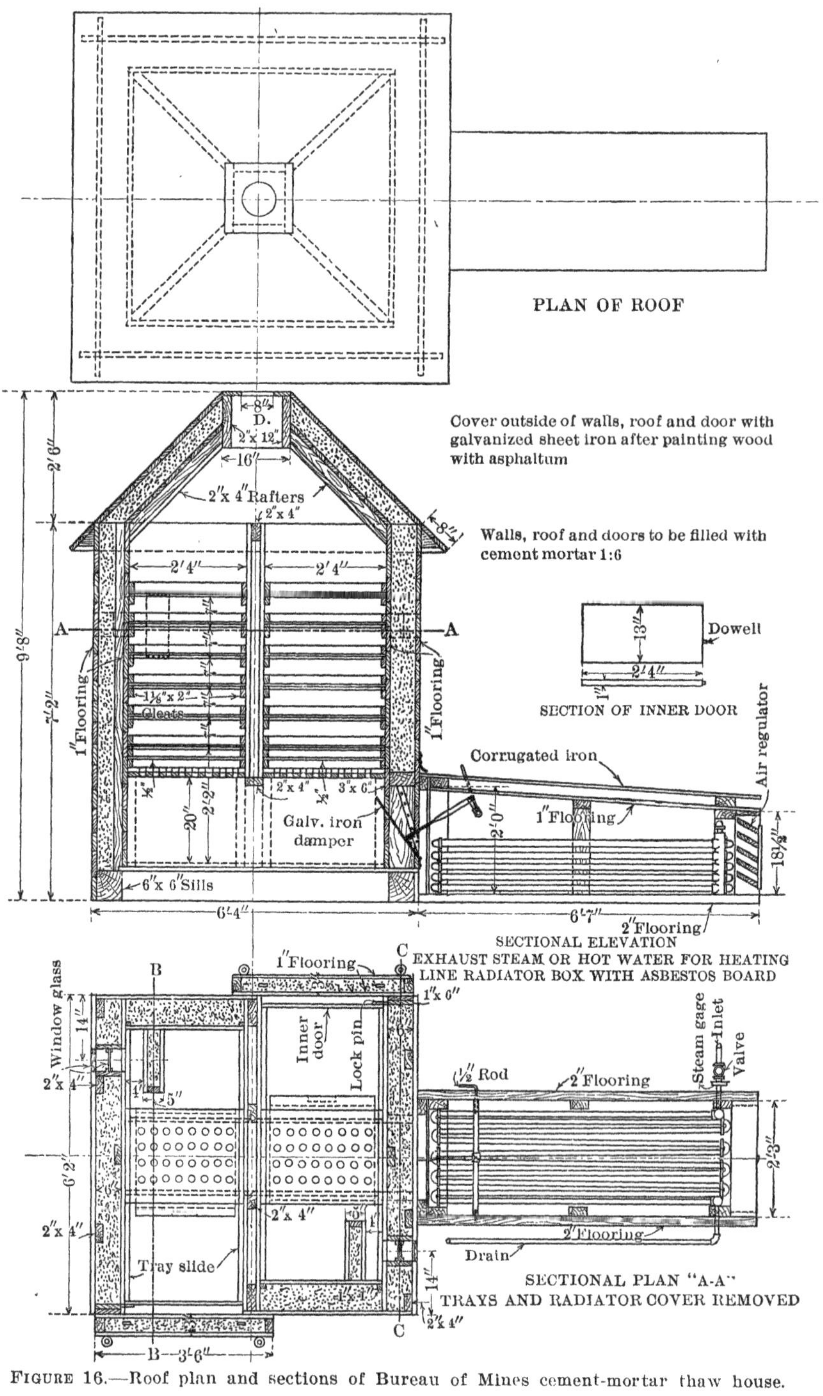
PLAN OF ROOF
Cover outside of walls, roof and door with galvanized sheet iron after painting wood with asphaltum
Walls, roof and doors to be filled with cement mortar 1:6
2″x 4″ Rafters
D.
2″x 12″
1″ Flooring
6″x 6″ Sills
Galv. iron damper
Corrugated iron
Air regulator
SECTION OF INNER DOOR
Dowell
SECTIONAL ELEVATION
EXHAUST STEAM OR HOT WATER FOR HEATING
LINE RADIATOR BOX WITH ASBESTOS BOARD
Window glass
Inner door
Lock pin
Tray slide
Steam gage
Inlet
Valve
Drain
2″ Flooring
SECTIONAL PLAN "A-A"
TRAYS AND RADIATOR COVER REMOVED

FIGURE 16.—Roof plan and sections of Bureau of Mines cement-mortar thaw house.

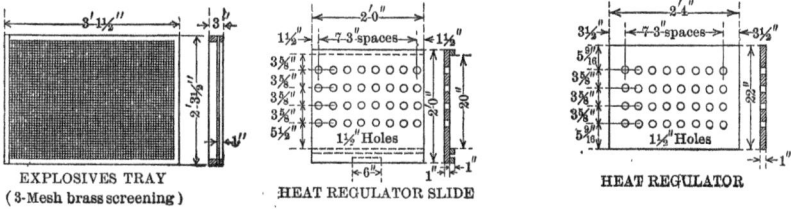

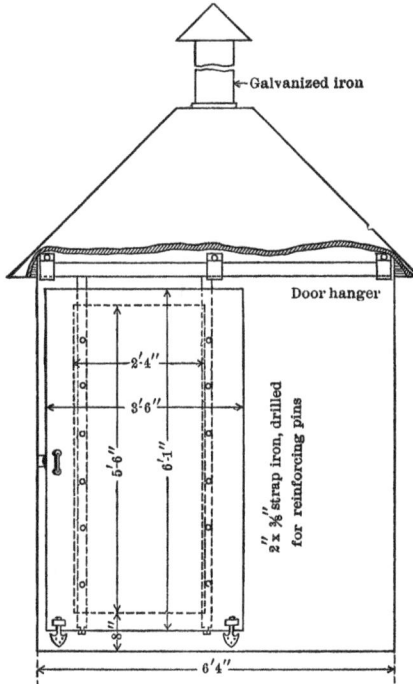

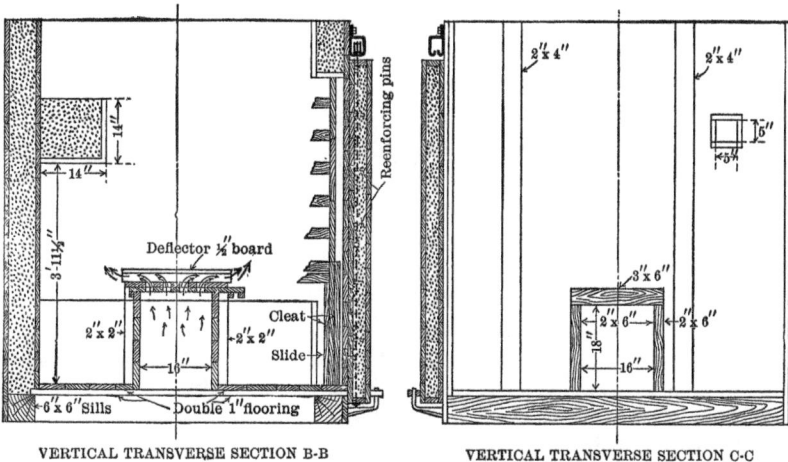

EXPLOSIVES TRAY
(3-Mesh brass screening)

HEAT REGULATOR SLIDE

HEAT REGULATOR

VERTICAL TRANSVERSE SECTION B-B

VERTICAL TRANSVERSE SECTION C-C

FIGURE 17.—Elevation and details of Bureau of Mines cement-mortar thaw house.

struction as a precautionary device for instantly cutting off the heating box from the air conduit and the explosives compartments.

Each compartment regulator consists of two 1-inch boards, the upper one stationary, the lower one sliding. Each board has 32 1½-inch holes, the holes in the movable board coinciding with those in the upper board. Each explosives compartment may then be supplied with equal quantities of air by partly closing the holes of the regulator for that compartment. Either compartment may be completely isolated from the incoming warm air by closing its air regulator, and hence one compartment may be used without the necessity of heating the other compartment, or the compartments may be used alternately. The steam coils consist of 50 lengths of 1-inch pipe, each 5 feet 2 inches long.

For the benefit of those who may be particularly interested in such data the following bill of materials used in the Bureau of Mines thaw house is presented:

BILL OF MATERIALS.

Lumber.

2 pieces of yellow-pine lumber 6 inches by 6 inches by 12 feet.
2 pieces of yellow-pine lumber 2 inches by 6 inches by 16 feet.
12 pieces of yellow-pine lumber 2 inches by 6 inches by 14 feet.
8 pieces of yellow-pine lumber 2 inches by 4 inches by 14 feet.
850 board feet of yellow-pine matched flooring 1 inch by 6 inches by 12 feet.
180 board feet of yellow-pine matched flooring 2 inches by 6 inches by 16 feet.
6 pieces of white pine 1¼ inches by 12 inches by 16 feet.
Cost, $40.26.

Hardware.

400 square feet of No. 24 galvanized flat iron.
3 pieces of galvanized flat iron 8 feet by 24 inches.
2 pieces of corrugated galvanized iron, standard 24-gage, 8 feet long.
45 1-inch return-pipe bends.
266 feet of 1-inch pipe.
2 1-inch plain, 5-branch tees.
94 square feet of brass wire cloth, 19-gage wire, 30 inches wide, 4 meshes per inch.
Nails.
2 pairs of No. 124 Coburn trolley hangers.
12 feet of No. 3 track.
4 end brackets and 2 center brackets.
Cost, $65.02.

Concrete.

5 barrels of cement.
4 cubic yards of sand.
2 cubic yards of gravel.
Cost, $10.61.
Total cost, $115.89.

HOT-WATER HEAT IN THAW HOUSES.

In some thaw houses it may be more convenient to use hot water as a means of heating. Hot-water heating requires a greater radiating surface than steam heating. If a thaw house must be isolated from any auxiliary source of steam or hot water, a small hot-water heater may be installed in accordance with the well-established principles of hot-water heating. The heater should, however, be installed in a separate compartment, placed at least 4 yards from the thaw house.

ELECTRIC HEATERS.

Electric heaters are known to be less efficient and less safe than low-pressure steam or hot-water coils for thawing explosives, nevertheless in certain cases electricity is the only available source of heat. With this consideration in view an investigation of the heating of thaw houses by electricity was undertaken by the bureau. The dangers encountered are stated as a warning in order to promote greater safety in thawing explosives when electric heating is the only alternative. In the tests the steam coils of the bureau's experimental thaw house were replaced with electrical means of heating. The conclusions drawn from the results of the tests are as follows:

FEATURES OF ELECTRIC HEATING.

Electricity will under certain conditions produce sparks and flashes sufficient to ignite explosives or to cause fires that may result in such ignition. It is therefore necessary to select and install heating devices, connections, and circuits with reference to the special service that they are to perform. By far the most unsatisfactory feature of electric heating is the probability of obtaining undesirably high temperatures close to the source of heat, and also the certainty that the temperature of such sources of heat may become dangerously high if the emission or diffusion of heat from them is sufficiently restricted.

COMPARISON OF HEATING METHODS.

Low-pressure steam or hot-water coils and electric heaters have exactly opposite characteristics. The rate at which steam coils give up heat is variable, depending on varying conditions, whereas the rate at which electric coils give up heat is constant. The temperature of steam coils can not exceed a certain maximum which depends upon the pressure of the steam used, regardless of radiation and convection. There is, however, almost no limit to the temperature that may be attained by an electric heater, and its maximum temperature depends entirely on those factors that affect the emission or diffusion of heat from the heaters. In thaw houses the most prominent factors are the weight of air passed per minute, its temperature, and its specific heat. The mechanical design of the heaters must also be considered.

From the foregoing it is evident that low-pressure or hot-water coils are better adapted for the warming of thaw houses, and that the operation of electrical heaters in connection with thaw houses requires much careful attention.

EFFECT OF HEAT CHANGES.

An increase in the temperature of the air surrounding a thaw house temporarily reduces the loss of heat from the house both by diminishing the amount of heat lost by radiation and convection and by lessening the amount of air circulating through the house. A reduction in the rate at which heat leaves the house must result in an increased temperature within the house. As electric heaters deliver a constant amount of heat, if the temperature outside rises, some of the heaters should be switched out of service (automatic switching by means of a thermostat being preferable) in order to keep the temperature inside from exceeding a safe limit. This limit might be exceeded if the temperature outside increased more than 10° F.

DESIRABLE CAPACITY OF HEATERS.

In the experimental thaw house of the bureau about 7 kilowatts of energy is required to maintain an average temperature of 90° F. within the house when the temperature of the air outside is 55° F. Therefore, heaters must be provided that have a maximum capacity of 7 kilowatts, a minimum capacity of 2 kilowatts, and several steps between these limits to take care of intermediate temperatures. A good combination is one heater of 2-kilowatt capacity and five heaters of 1-kilowatt capacity each.

MISCELLANEOUS ESSENTIALS OF HEATERS.

The compartment in which the heaters are installed should be built entirely of noncombustible material. Each heater should be provided with a switch and fuses placed on the outside of the heater compartment. The heater should be fused as low as possible, and both switches and fuses should be carefully installed so as to be protected from injury. The heaters should be designed with large radiating surfaces so that the difference between their temperatures and the average temperature of the air in the heating compartment may be as small as possible. It is recommended that the heaters be composed of units, each unit consisting of a single layer of uninsulated resistance wire wound on a porcelain tube. The wire should be wound in a spiral groove or should be covered with an insulating enamel or equivalent substance. The units should be mounted on an iron framework, from which they should be thoroughly insulated. There should be as few connections about them as possible, and these should be made with great thoroughness. If solder is used, its melting point should be not less than 300° F. Only flame-proof insulated wire should be used for the leads or for any connection. The voltage supply should not be higher than 250 volts, and preferably not higher than 110 volts.

In thaw houses heated by electricity there should be several thermometers, one of which should be in the heating box between the heaters and the opening through which the heated air enters the house.

It is advised that electricity should not be used for thawing explosives unless the foregoing suggestions can be followed and unless careful attention is given to the operation of the electrically-heated thaw house while it is in service.

PERMISSIBLE EXPLOSIVES.

PERMISSIBLE EXPLOSIVES FOR COAL MINES.

From the foundation of the Pittsburgh testing station that branch of the Bureau of Mines has, among other duties, been engaged in the testing of explosives proposed for use in coal mines. A definitely prescribed system of tests has been used, and those explosives passing the tests have been styled "permissible explosives," and because of the increase in safety which results from their substitution for the explosives previously employed in coal mining their use has since 1909 increased with each succeeding year.

PROPOSED REQUIREMENTS FOR METAL-MINE EXPLOSIVES.

There is rarely in metal mining and in quarrying the menace of combustible dust and gas that is encountered in coal mining, yet it is believed that safety and efficiency in metal mining and quarrying would be improved if the explosives offered for use in those industries were subjected to prescribed official tests also. The following proposed requirements of explosives for metal mines are offered tentatively:

1. The explosives must be in a stable condition and none of the chemical and physical tests made upon them shall show any unfavorable results.

2. The explosives, when detonated in 200-gram charges in the pressure gage, must not evolve carbon monoxide, hydrogen sulphide, nitrogen oxides, or other permanent poisonous gases in quantities that may be considered harmful to the health of miners. ·

3. Such electric detonators will be used in the tests as are recommended by the manufacturers, provided the grade recommended completely detonates or explodes the charge.

4. The explosives must pass friction-pendulum tests in which the steel shoe is faced with wood fiber.

5. The propulsive and disruptive effects developed by each explosive shall be determined and the results published in order to serve as a useful guide in indicating the efficiency of the explosive.

6. After an explosive shall have passed the above requirements it

112

will be considered permissible for use only when used under the following conditions:

(*a*) That the explosive is in all respects similar to the sample submitted by the manufacturer for test.

(*b*) That detonators, or electric detonators, are used of not less efficiency than those prescribed, namely, those consisting by weight of 90 parts of mercury fulminate and 10 parts of potassium chlorate (or their equivalents).

(*c*) That the explosive, if frozen, before use, shall be thoroughly thawed in a safe and suitable manner.

62626°—Bull. 80—15——8

REQUIREMENTS FOR BUREAU OF MINES TESTS OF METAL-MINE EXPLOSIVES.

The Bureau of Mines has prescribed the following conditions and requirements in regard to the testing of explosives for use in metal mines, tunnels, quarries, and other engineering operations.[a]

TESTS OF EXPLOSIVES USED IN METAL MINES, TUNNELS, QUARRIES, AND OTHER ENGINEERING OPERATIONS.

FEES.

For a complete official test of each explosive to determine its suitability for use in metal mines, tunnels, quarries, and other engineering operations__ $50

For two experimental shots in ballistic pendulum and two in gage to determine propulsive effect_____ 20

For two experimental shots in Trauzl lead blocks, two on small lead blocks, and two rate of detonation (Mettegang's recorder) tests, to determine disruptive effect _____ 12

For one gage test to determine the products of combustion_____ 6

For pendulum-friction test to determine sensitiveness to a glancing blow___ 5

For impact test to determine sensitiveness to direct impact_____ 8

For calorimeter test to determine calories developed_____ 6

CONDITIONS UNDER WHICH TESTS WILL BE MADE.

The conditions under which the Bureau of Mines will test explosives to determine their suitability for use in metal mines, tunnels, quarries, and other engineering operations are as follows:

1. The manufacturer or applicant desiring tests to be made is to deliver to the Bureau of Mines, Fortieth and Butler Streets, Pittsburgh, Pa., three weeks prior to date set for tests, 50 pounds [b] of each explosive that he desires to have tested. He is to be responsible for the care, handling, and delivery of this material to the experiment station, and, if he desires, he may have a representative present during the tests.

2. No one is to be present at or participate in these tests except the necessary Government officers at the experiment station, their assistants, the representative of the manufacturer of the explosives, or the applicant desiring tests to be made.

3. These tests will be made in the order of the receipt of the applications for them, provided the necessary quantity of the explosive is delivered at the experiment station by the date set, of which date due notice will be given by the Bureau of Mines.

4. A list of the explosives that pass certain requirements satisfactorily will be furnished to the State mine inspectors in the several States and will be made public in such manner as may be considered desirable.

5. The details of results of tests are to be considered confidential and are not to be made public prior to official publication by the Bureau of Mines.

---

[a] Fees for testing explosives, and conditions and requirements under which explosives are tested, Schedule 1 (approved Sept. 17, 1913), Bureau of Mines, 1913, pp. 7-8.

These requirements are subject to revision and may be changed this year (1915).

[b] This requirement has been changed so that the manufacturer submits only the necessary quantity.

6. From time to time field samples of explosives will be collected, and tests will be made of these explosives as they are supplied for use. Field samples collected in original shipping case will be tested and analyzed. The proportion of each ingredient of the explosive must agree, within a reasonable variation, with that in the original sample submitted for tests.

TEST REQUIREMENTS.

Each charge shall be fired with an electric detonator (exploder or cap)[a] of not less efficiency than a No. 6 or its equivalent.

The explosive must be in such condition that the chemical and physical tests do not show any unfavorable results. An explosive will be considered unsatisfactory if it is not chemically stable, if it shows leakage of nitroglycerin, or if it is in such condition that exudation of nitroglycerin would occur in handling or transportation.

An explosive will be considered unsatisfactory for use in metal mines, tunnels, and similar operations if, on detonation in the gages, it evolves poisonous gases in quantities that may be considered harmful to the health of the miners.

An explosive will be considered unsatisfactory if it is too sensitive to frictional impact (a glancing blow). This sensitiveness will be determined with the pendulum-friction device with the shoe faced with fiber.

An explosive will be considered unsatisfactory if in the course of the official tests two or more charges fail to detonate or explode completely, or if after two months' storage at the Pittsburgh experiment station two or more cartridges fail to detonate or explode completely when fired separately with a suitable detonator.

REMITTANCES.

Applicants who submit explosives for tests will be required to furnish a certified check or bank draft, made payable to the Secretary of the Interior, to cover the total fee required for the tests desired. Such fees must be received at least three weeks prior to the date set for beginning the test.

J. A. HOLMES, *Director.*

Approved, September 17, 1913.

ANDRIEUS A. JONES, *Acting Secretary.*

---

[a] Electric igniters will be used with slow-burning explosives.

SAFE SHIPMENT AND STORAGE OF EXPLOSIVES.

By B. W. DUNN.

RESPONSIBILITY OF MANUFACTURERS AND COMMON CARRIERS TO PUBLIC.

A responsibility to the public rests upon both manufacturers and common carriers to secure the safe delivery at destination of explosives, and it is the duty of the owners of explosives to store them safely.

FEDERAL LAW AND INTERSTATE COMMERCE COMMISSION REGULATIONS.

Under authority granted by Congress, the Interstate Commerce Commission has made regulations,[a] binding upon shippers and common carriers, for the transportation of explosives in interstate commerce, and the penalty of a possible fine of $2,000 and 18 months' imprisonment is prescribed by law for a violation of any of these regulations.

The shipper must know and certify on his shipping order that the explosive offered by him is in a proper condition for safe transportation and that it is packed and marked as required by the regulations. To perform this duty the shipper should be thoroughly familiar with all requirements pertaining to his shipment. A copy of the regulations can be obtained by application to the railway agent, whose duty it is to furnish them to shippers.

The following paragraphs in these regulations are of special interest to the shippers of explosives used in mines: *General Rule A.*— Paragraphs 1501, 1502, 1503, 1509, 1510, 1531, 1533, 1541 to 1556, 1558 to 1560, 1611 to 1614, 1648, 1661, 1665, 1666, 1668, 1674 to 1683.

EXPLOSIVES IN BAGGAGE OR HOUSEHOLD GOODS.

Miners and other persons are sometimes tempted to pack explosives for shipment with their baggage on passenger cars or with their household goods for shipment by freight. To do this is a criminal act that may endanger the lives of the innocent and unsus-

---

[a] Regulations for the transportation of explosives and other dangerous articles by freight and express, and specifications for shipping containers, 1914, p. 196.

pecting persons who have to handle these packages, and will subject the guilty shipper, when detected, to arrest and prosecution. The Federal law (sec. 236) prescribes an imprisonment of 10 years for anyone convicted of this crime when death or bodily injury results from the illegal transportation of explosives. When no injury results, the maximum penalty is 18 months' imprisonment and a fine of $2,000.

Persons receiving packages of explosives sent by rail should examine them carefully to discover ruptures or other serious damage during transit. Any information regarding such matters will be welcomed by the Chief Inspector, Bureau for the Safe Transportation of Explosives, 24 Park Place, New York City.

PUBLICATIONS ON MINE ACCIDENTS AND TESTS OF EXPLOSIVES.

A limited supply of the following publications of the Bureau of Mines is temporarily available for free distribution. Requests for all publications can not be granted, and applicants should limit their selection to publications that may be of especial interest to them. Requests for publications should be addressed to the Director, Bureau of Mines, Washington, D. C.

BULLETIN 15. Investigations of explosives used in coal mines, by Clarence Hall, W. O. Snelling, and S. P. Howell, with a chapter on the natural gas used at Pittsburgh, by G. A. Burrell, and an introduction by C. E. Munroe. 1911. 197 pp., 7 pls., 5 figs.

BULLETIN 17. A primer on explosives for coal miners, by C. E. Munroe and Clarence Hall. 61 pp., 10 pls., 12 figs. Reprint of United States Geological Survey Bulletin 423.

BULLETIN 20. The explosibility of coal dust, by G. S. Rice, with chapters by J. C. W. Frazer, Axel Larsen, Frank Haas, and Carl Scholz. 204 pp., 14 pls., 28 figs. Reprint of United States Geological Survey Bulletin 425.

BULLETIN 42. The sampling and examination of mine gases and natural gas, by G. A. Burrell and F. M. Seibert. 1913. 116 pp., 2 pls., 23 figs.

BULLETIN 46. An investigation of explosion-proof mine motors, by H. H. Clark. 1912. 44 pp., 6 pls., 14 figs.

BULLETIN 48. The selection of explosives used in engineering and mining operations, by Clarence Hall and S. P. Howell. 1913. 50 pp., 3 pls., 7 figs.

BULLETIN 52. Ignition of mine gases by the filaments of incandescent electric lamps, by H. H. Clark and L. C. Ilsley. 1913. 31 pp., 6 pls., 2 figs.

BULLETIN 56. First series of coal-dust explosion tests in the experimental mine, by G. S. Rice, L. M. Jones, J. K. Clement, and W. L. Egy. 1913. 115 pp., 12 pls., 28 figs.

BULLETIN 59. Investigations of detonators and electric detonators, by Clarence Hall and S. P. Howell. 1913. 73 pp., 7 pls., 5 figs.

BULLETIN 66. Tests of permissible explosives, by Clarence Hall and S. P. Howell. 1913. 313 pp., 1 pl., 6 figs.

BULLETIN 68. Electric switches for use in gaseous mines, by H. H. Clark and R. W. Crocker. 1913. 40 pp., 6 pls.

BULLETIN 69. Coal-mine accidents in the United States and foreign countries, compiled by F. W. Horton. 1913. 102 pp., 3 pls., 40 figs.

BULLETIN 82. International conference of mine-experiment stations. Pittsburgh. Pa., September 14–21, 1912, compiled by G. S. Rice. 1914. 99 pp., 4 figs.

TECHNICAL PAPER 6. The rate of burning of fuse as influenced by temperature and pressure, by W. O. Snelling and W. C. Cope. 1912. 28 pp.

TECHNICAL PAPER 7. Investigations of fuse and miners' squibs, by Clarence Hall and S. P. Howell. 1912. 19 pp.

TECHNICAL PAPER 11. The use of mice and birds for detecting carbon monoxide after mine fires and explosions, by G. A. Burrell. 1912. 15 pp.

TECHNICAL PAPER 12. The behavior of nitroglycerin when heated, by W. O. Snelling and C. G. Storm. 1912. 14 pp., 1 pl., 2 figs.

TECHNICAL PAPER 13. Gas analysis as an aid in fighting mine fires, by G. A. Burrell and F. M. Seibert. 1912. 16 pp., 1 fig.

TECHNICAL PAPER 17. The effect of stemming on the efficiency of explosives, by W. O. Snelling and Clarence Hall. 1912. 20 pp., 11 figs.

TECHNICAL PAPER 18. Magazines and thaw houses for explosives, by Clarence Hall and S. P. Howell. 1912. 34 pp., 1 pl., 5 figs.

TECHNICAL PAPER 21. The prevention of mine explosions, report and recommendations, by Victor Watteyne, Carl Meissner, and Arthur Desborough. 12 pp., Reprint of United States Geological Survey Bulletin 369.

TECHNICAL PAPER 30. Mine-accident prevention at Lake Superior iron mines, by D. E. Woodbridge. 1913. 38 pp., 9 figs.

TECHNICAL PAPER 39. The inflammable gases in mine air, by G. A. Burrell and F. M. Seibert. 1913. 24 pp., 2 figs.

TECHNICAL PAPER 40. Metal-mine accidents in the United States during the calendar year 1911, compiled by A. H. Fay. 1913. 54 pp.

TECHNICAL PAPER 41. Mining and treatment of lead and zinc ores. Joplin district, Missouri; a preliminary report, by C. A. Wright. 1913. 43 pp., 5 figs.

TECHNICAL PAPER 43. The influence of inert gases on inflammable gaseous mixtures, by J. K. Clement. 1913. 24 pp., 1 pl., 8 figs.

TECHNICAL PAPER 44. Safety electric switches for mines, by H. H. Clark. 1913. 8 pp.

TECHNICAL PAPER 46. Quarry accidents in the United States during the calendar year 1911, compiled by A. H. Fay. 1913. 32 pp.

TECHNICAL PAPER 47. Portable electric mine lamps, by H. H. Clark. 1913. 13 pp.

TECHNICAL PAPER 48. Coal-mine accidents in the United States, 1896–1912, with monthly statistics for 1912, compiled by F. W. Horton. 1913. 74 pp., 10 figs.

TECHNICAL PAPER 52. Permissible explosives tested prior to March 1, 1913, by Clarence Hall. 1913. 11 pp.

TECHNICAL PAPER 59. Fires in Lake Superior iron mines, by Edwin Higgins. 1913. 34 pp., 2 pls.

TECHNICAL PAPER 61. Metal-mine accidents in the United States during the calendar year 1912, compiled by A. H. Fay. 1913. 76 pp., 1 fig.

TECHNICAL PAPER 62. Relative effects of carbon monoxide on small animals, by G. A. Burrell, F. M. Seibert, and I. W. Robertson. 1914. 23 pp.

TECHNICAL PAPER 67. Mine signboards, by Edwin Higgins and Edward Steidle. 1913. 15 pp., 1 pl., 4 figs.

TECHNICAL PAPER 69. Production of explosives in the United States in the calendar year 1912, compiled by A. H. Fay. 8 pp.

TECHNICAL PAPER 71. Permissible explosives tested prior to January 1, 1914, by Clarence Hall. 1914. 12 pp.

MINERS' CIRCULAR 5. Electrical accidents in mines, their causes and prevention, by H. H. Clark, W. D. Roberts, L. C. Ilsley, and H. F. Randolph. 1911. 10 pp., 3 pls.

MINERS' CIRCULAR 6. Permissible explosives tested prior to January 1, 1912, and precautions to be taken in their use, by Clarence Hall. 1912. 20 pp.

MINERS' CIRCULAR 7. Use and misuse of explosives in coal mining, by J. J. Rutledge, with a preface by J. A. Holmes. 1913. 52 pp., 8 figs.

MINERS' CIRCULAR 8. First-aid instructions for miners, by M. W. Glasgow, W. A. Raudenbush, and C. O. Roberts. 1913. 66 pp., 46 figs.

MINERS' CIRCULAR 10. Mine fires and how to fight them, by J. W. Paul. 1912. 14 pp.

MINERS' CIRCULAR 12. The use and care of miners' safety lamps, by J. W. Paul. 1913. 16 pp., 4 figs.

MINERS' CIRCULAR 13. Safety in tunneling, by D. W. Brunton and J. A. Davis. 1913. 19 pp.

MINERS' CIRCULAR 14. Gases found in coal mines, by G. A. Burrell. 1913. 21 pp.

MINERS' CIRCULAR 15. Rules for mine-rescue and first-aid field contests, by J. W. Paul. 1913. 12 pp.

MINERS' CIRCULAR 16. Hints on coal-mine ventilation, by J. J. Rutledge. 1914. 22 pp.

MINERS' CIRCULAR 17. Accidents in metal mines from falls of roof, by Edwin Higgins. 1914. 15 pp., 8 figs.

MINERS' CIRCULAR 19. The prevention of accidents from explosives in metal mining, by Edwin Higgins. 1914. 16 pp., 11 figs.

INDEX.

A.

Page.

"Adobe" work on bowlders, explosive
for _____ 55
Alcohol, use of, in making explosive
    . compounds_____ 12
Ammonia dynamites, composition of_ 23
    for excavation work_____ 55
Ancon quarry, Canal Zone, blasting
        in, cost of_____ 69
        explosives used in_____ 69
        method of _____ 69
        extent of_____ 69
        face of, view of_____ 40
Austria, required distance of maga-
        zines from dwellings in_ 99

B.

Ballistic pendulum, view of_____ 46
Ballistic-pendulum tests, details of__ 45
    results of_____ 46
Barricades for magazines, construc-
        tion of_____ 99
Batteries for firing charges, testing
        strength of _____ 50
    See also Dry-cell batteries.
Berthelot, on gases from nitroglyc-
        erin_____ 19
Bichel pressure gage, details of_____ 24, 25
    view of_____ 16
Bickford, William, invention of fuse
        by_____ 29
Black powder, characteristics of____ 18, 19
    composition of _____ 19
    firing of_____ 47
    gases from_____ 19, 25
    grades of_____ 19
    making of _____ 18, 19
    rate of burning_____ 20
    uses of_____ 20, 21
Blast, large, cost of_____ 57, 58
    description of_____ 55-58
    details of_____ 58
    explosive used in_____ 55
    gases from_____ 59
    placing of charge for_____ 57
        figure showing_____ 57
    results of_____ 59
Blasting, electricity for, sources of_ 49
Blasting caps., *See* Detonators.
Blasting machines. *See* Firing ma-
        chines.

Page.

Blasting powder. *See* Black powder.
Blasting, submarine, procedure in__ 60
    selection of explosives for_ 18,
        24, 59-60
"Block holing," explosive for_____ 55
Boosters, use of_____ 35
Bore holes, connecting of, in series_ 48, 49
"Brisance," definition of _____ 14
Buffalo waterworks tunnel, arrange-
        ment of drill holes in,
        figure showing _____ 76
    blasting in, detonators used in_ 77
    explosives used in _____ 76
        handling of_____ 77
    features of _____ 75, 76
    firing of shots in_____ 77
    method of driving _____ 76
    record of drilling_____ 86
"Bulldozing." *See* "Adobe" work.
Bureau of Mines, investigations of__ 9,
        26, 96, 112
    results of_____ 42
Bureau of Mines tests of explo-
        sives, conditions govern-
        ing _____ 114
    fees for _____ 114
    requirements for _____ 115
Burning. *See* Combustion.

C.

Canal Zone, fuse used at, require-
        ments for _____ 31
    quarries in. *See* Porto Bello
        quarry; Ancon quarry.
Carbon monoxide, production of, by
        explosives _____ 25
Chicago waterworks tunnel, record
        of drilling_____ 86
Cities, blasts in, covers for_____ 61
    explosives used in _____ 61
Clip for detonators _____ 39
Coal mines, combination charges in,
        danger from_____ 45, 46
    permissible explosives for_____ 112
    selection of explosives for _____ 18
Combustion, nature of_____ 10
Copper, for leading wires, advantages
        of _____ 52
Cordeau Bickford, characteristics of_ 35
Cordeau detonant, characteristics of_ 35
    use of_____ 34, 64

Page.

Crarae quarry, Scotland, explosion
  at_____ 16
  view of_____ 16
Crimper, features of_____ 75
  operation of_____ 74
  views of_____ 74

D.

Delay-action detonators. *See* Detonators.
Detonation, definition of_____ 13
  explosion by, advantage of____ 14
  method of_____ 13
  rate of_____ 20
  See also various explosives
  named.
Detonator, crimping of, view of___ 38
  advantages of high-grade_____ 42
Detonators, box for, view of_____ 40
  clip for, description of_____ 39
    view of_____ 40
  delay-action, for blasting_____ 73, 77
  details of_____ 35, 36
  distribution of, in charge_____ 48
  firing of_____ 37
  fused, difficulty of using, in wet
    blasting_____ 74
  grades of_____ 36
  low-grade, disadvantage of____ 42
  storage of_____ 40
  transportation of_____ 40
  use of, advantages of_____ 36
    precautions in_____ 39
    view of_____ 38
  See also Electric detonators.
Disruptive force. *See* various explosives named.
Drill holes, chambering of, value of_ 21
  cooling effect of_____ 20
Dry-cell batteries for firing charges,
    danger from_____ 49
    safety spring for_____ 49
Dusts, inflammable, as cause of ex-
  plosions_____ 10
Dynamites. *See* Ammonia dyna-
    mites; gelatin dyna-
    mites; low-freezing dyna-
    mites; nitroglycerin dy-
    namites.

E.

Electric detonators, advantage of___ 72, 77
  clip for, description of_____ 39
    view of_____ 40
  connecting of, in series_____ 48
  delay-action, description of____ 39
  description of_____ 37, 38
  distributed in charge, tests of,
    results of_____ 42, 43
  figure showing_____ 37
  firing of_____ 38, 39
  grades of_____ 38
  loading of_____ 47

Page.

Electric detonators, nail test for___ 40, 41
  results of, view of_____ 42
  simultaneous firing of_____ 44
  view of _____ 38
  waterproof, for submarine blast-
    ing _____ 60
Electric heating, for thaw houses,
    dangers of_____ 109
    features of _____ 109
Electric igniters, description of____ 46
  loading of, into bore hole_____ 47
Electric-light circuit, firing from___ 50, 51
Elizabeth Tunnel, Cal., arrangement
    of drill holes in figure
    showing_____ 81
  driving of_____ 81, 82
    explosives used in_____ 81
  features of _____ 80, 81
  record of drilling_____ 86
England, magazines in, required dis-
    tance of, from dwellings_ 99
  use of lightning conductors
    on_____ 96
Excavations, selection of explosives
    for _____ 54
    factors governing _____ 54, 55
Exploders. *See* Detonators.
Explosion, causes of_____ 10
  description of_____ 10
  pressure exerted by_____ 14
  products of, determination of__ 24, 25
Explosives, accidents from, number
    injured in_____ 9
    number killed in_____ 9
  classification of_____ 14, 17
  combination charges of, advan-
    tage of_____ 44
    danger from_____ 45
  devices for firing of_____ 49–52
  efficiency of, factors governing_ 14,
                                    19, 20
  explosion of, methods for_____ 29
  for coal mining_____ 9
  for metal mining_____ 9
  insensitive, increase of efficiency
    of _____ 42
  manufacture of oxidizing agents
    used in_____ 11
  necessity of confining_____ 15
  packages of, opening of_____ 95
  pressures developed by, deter-
    mination of_____ 24, 25
  selection of_____ 18
  storage of_____ 94
  tests of, by Bureau of Mines,
    conditions governing_____ 114
    fees for_____ 114
    requirements for_____ 115
  thawing of. *See* Thawing.
  transportation of, penalty for
    illegal _____ 117
    regulations regarding_____ 116
  variations in _____ 17
  See also High explosives; low
    explosives; and various
    explosions named.

F.

Page.

Fees for Bureau of Mines tests of
    explosives _____ 114
Firing machines, details of_____ 51, 52
    determination of capacity of___ 50
    leading wires for_____ 52
    proper manipulation of_____ 50, 51
    rating of_____ 52
    types of_____ 51, 52
    use of _____ 49
    view of _____ 46
Franklin Furnace, N. J., mine at, ex-
    periments at_____ 28
Fuse, burning, action of_____ 30
    advantages of_____ 29, 30
    attaching of, to detonators_ 37
    burning of, rate of_____ 30
        need of showing_____ 33
        testing of_____ 31
    classes of_____ 29, 32, 33
        characteristics of_____ 33
    cutting of _____ 33, 34
    for detonators_____ 32
    invention of_____ 29
    kinds of_____ 30
    requirements for_____ 31
    storage of_____ 32
    transportation of_____ 32
    view of _____ 30
    defective, side splitting of,
        view of_____ 32
        view of_____ 32
    detonating, description of_____ 34, 35

G.

Galvanometer, testing of firing line
    with _____ 52, 53
Gases, combustible, as cause of explo-
    sions_____ 10
    from explosion, danger from___ 15
    production of, prevention of_ 15
Gelatin dynamite, approved, tests of_ 27, 28
        results of_____ 28
    composition of _____ 23
    for wet blasting, value of_____ 24, 60
    poisonous gases from_____ 25
    conditions governing production
        of _____ 28
    fatalities from _____ 25, 26
    well-balanced formula for_____ 27
Germany, magazines in, lightning con-
    ductors on_____ 96
    required distance of, from
        dwellings_____ 99
Granulated powder. See Nitroglycerin
    powder, granulated.
Guncotton. See Nitrocellulose.
Gunnison Tunnel, Colo., record of
    drilling_____ 86
Gunpowder, origin of_____ 11

H.

Hazlewood, A. J., on blasting at
    Tomkins Cove, N. Y____ 70

Page.

High explosives, characteristic of__ 17
    explosion of, view of_____ 14
Hot-water heat for thaw houses___ 109
Hunter Brook Tunnel, N. Y., driv-
    ing of_____ 79, 80
    explosives used in_____ 80
    figure showing _____ 80
    features of _____ 79, 80
    record of drilling_____ 86
Hustvedt, O. M., on blasting in
    Pittsburgh, Pa_____ 61, 62
Hydrogen sulphide, production of,
    by explosives _____ 25

I.

Institute of Makers of Explosives, on
    proper strength of deto-
    nators_____ 36, 37
Interstate Commerce Commission,
    regulations of, on trans-
    porting explosives _____ 116
Italy, magazines in, required dis-
    tance of, from dwellings_ 99

K.

Kellogg Tunnel, Idaho, record of
    drilling_____ 86

L.

Laramie-Poudre Tunnel, Colo., ar-
    rangement of drill holes
    in, figure showing_____ 83, 84
    drilling record of_____ 85, 86
    driving of, fuse used in_____ 84
    explosives used in_____ 83, 84
        tests of _____ 84
        ventilation during_____ 85
    features of _____ 82, 83
Leading wires, description of_____ 52
    need of inspecting_____ 52
    repairing of_____ 52
Loetschberg Tunnel, driving of, de-
    tails regarding___ 89, 90, 92, 93
    explosives used in_____ 90
    features of_____ 89
    sections of, figure showing____ 89
Los Angeles Aqueduct, Cal., record
    of drilling_____ 86
Low explosives, characteristic of___ 17
Low-freezing dynamites, compo-
    sition of_____ 23
    poisonous gases from_____ 25

M.

Magazine, barracades for, construc-
    tion of_____ 99
    bullet-proof, need of_____ 96
    Bureau of Mines, construction
        of _____ 96, 98
        features of_____ 100
        materials used in__ 100, 101

Page.

Magazine, Bureau of Mines, merits
of_____ 100
   sections of figure showing_ 97
   view of_____ 96
   construction of, materials for__ 96,
                                99, 100
   lightning conductors for_____ 96
   need of _____ 94
   permitted distance of, from
      dwellings_____ 98, 99
   proper care of_____ 95, 96
   proper site for_____ 94, 98
   repair of_____ 95
   storage of explosives in_____ 95
   supervision of explosives in___ 95
Manure, thawing of explosives by- 102
Massachusetts, magazines in, re-
      quired minimum dis-
      tances of, from dwell-
      ings _____ 99
Mercury fulminate, method of con-
      fining in detonator_____ 40
Metal mines, blasting in shafts of_ 72–74
   explosives used in, proposed re-
      quirements for_____ 112, 113
   selection of factors govern-
      ing _____ 18
   features of_____ 72
Morse, C. W., on use of delay-action
      detonators_____ 73

N.

Nail test for detonators, details of__ 40, 41
   figure showing_____ 41
   results of _____ 42
New York rapid-transit tunnel, drill
      holes in, details of_____ 78
   method of excavating_____ 78
Nitric acid, use of, as oxidizing
      agent _____ 11
Nitric esters, classification of_____ 12
Nitrocellulose, manufacture of_____ 11, 12
Nitrogen compounds, explosive, char-
      acteristics of _____ 13
Nitroglycerin, making of_____ 12
   volume of gases from_____ 19
Nitroglycerin dynamite, advantages
      of, for submarine blast-
      ing _____ 60
   composition of _____ 22
   disruptive force of_____ 22
   formulas for _____ 26
   poisonous gases from _____ 25
   rate of detonation of_____ 22
Nitroglycerin powder, granulated,
      composition of _____ 21
   method of detonating_____ 21
   poisonous gases from _____ 25
   rate of detonation of _____ 21
Nitrosubstitution compounds, classi-
      fication of, factors deter-
      mining _____ 12
   making of, conditions governing_ 12

O.
Page.

Ouray Tunnel, Colo., record of
      drilling_____ 86
Oxygen in explosives, sources of____ 11
   need of, for combustion _____ 10, 12

P.

Permissible explosives, definition of_ 112
Pittsburgh, Pa., blasting building site
      in _____ 61, 62
"Plastering" work on bowlders, se-
      lection of explosive for__ 55
Porto Bello quarry, blasting in, cost
      of _____ 69
   explosives used in _____ 68
   method of _____ 68
   excavations in, extent of _____ 68
Potassium chlorate, use of, as oxidiz-
      ing agent_____ 11
Potassium nitrate. See Saltpeter.

Q.

Quarrying, bench method of_____ 63
   explosives used in _____ 18
   selection of_____ 63
   face method of _____ 64

R.

Railroad cut, blasting in, cost of____ 57, 58
   description of _____ 55–58
   details of_____ 58
   explosives used in _____ 55
   figure showing _____ 56
   method of _____ 56
   placing of charge for_____ 57
   figure showing_____ 57
   results of_____ 59
Rate-of-detonation tests, results of__ 43
Reclamation Service, fuse used by,
      requirements for_____ 31
Renforts. See Boosters.
Rondout Tunnel, N. Y., arrangement
      of drill holes in, figure
      showing _____ 79
   excavation of, explosives used
      in _____ 79
   method of _____ 78, 79
   features of _____ 78, 79
   record of drilling_____ 86
"Running" fuse. See Fuse, burn-
      ing.

S.

"Safety" fuse. See Fuse, burning.
St. Pauls Pass Tunnel, drilling rec-
      ord at _____ 86
Saltpeter, use of, as oxidizing agent_ 11
Shafts in metal mines, blasting in,
      dangers of_____ 72
   detonators used for_____ 72, 73

Page.

Sheet iron, galvanized, use of for
   magazines ---------------- 96, 98
Simplon Tunnel, explosives used in_ 88
   features of ------------------ 87, 88
   method of driving ------------ 88
   method of ventilating-------- 88, 89
   sections of, figure showing----- 87
Sodium nitrate. *See* Saltpeter.
Souder, Harrison, on cordeau detonant -------------------- 64
Squib, view of--------------------- 30
Stemming, definition of------------ 15
   importance of---------------- 20
Submarine blasting. *See* Blasting.

T.

Tallow, sealing detonators with---- 75
Tarapaca, Chile, saltpeter found in_ 11
Tenino, Wash., quarry at, charges
   in, figure showing------ 66
   description of------------ 65
   explosives used in-------- 65
   result of blast at--------- 67
   system of wiring at------- 65, 67
   figure showing------- 67
Thaw house, Bureau of Mines, details of, figure showing_ 106
   features of-------------- 107
   materials used in---------- 108
   merits of --------------- 107
   operation of------------- 107
   sections of, figure showing_ 105
   view of------------------ 96
   construction of, precautions in_ 104
   heaters for, essential features
   of ------------------ 110, 111
   manure, features of--------- 103, 104
Thaw house, methods of heating__ 104, 109
   comparison of----------- 109
   within mine, danger from----- 103

Page.

Thawed explosives, characteristics
   of --------------------- 102
   transporting of, precautions in_ 103
Thawing of explosives, devices for,
   details of ------------- 102
   view of----------------- 96
   methods for ---------------- 102, 103
   need for----------------- 101
   precautions in-------------- 101, 102
Tomkins Cove, N. Y., quarry at, explosives used in-------- 70, 71
   method of using--------- 70
Trans-Andine Summit Tunnel, South
   America, driving of, explosive used in--------- 86, 87
   method of--------------- 86, 87
   features of------------------ 86
Tunneling, explosives used in------ 24
   desirable characteristics of_ 18
Tunnels, method of driving-------- 73, 74
   See also tunnels named.

U.

United States, magazines in, proposed distances of, from
   dwellings--------------- 99
Utah, method of quarrying in------ 64

V.

Ventilation of tunnels. *See* various
   tunnels named.

W.

West Winfield, Pa., mine at, experiment at------------ 27
Wet blasting. *See* Blasting, submarine.

www.ingramcontent.com/pod-product-compliance
Lightning Source LLC
Chambersburg PA
CBHW031406180326
41458CB00043B/6628/J

.